UTSA
Memorial Fund
in memory of

Ed Wagenfuehr

From a Gift
to UTSA by

Dean James F. Gaertner
College of Business

FRONTIERS OF SPACE

Herbert Friedman, General Editor

FRONTIERS OF SPACE

Rockets into Space   Frank H. Winter

Planet Earth: The View from Space   D. James Baker

Space Commerce   John L. McLucas

SPACE COMMERCE

JOHN L. McLUCAS

Harvard University Press · Cambridge, Massachusetts · London, England · 1991

This book is printed on acid-free paper, and its binding materials
have been chosen for strength and durability.

Library of Congress Cataloging-in-Publication Data
McLucas, John L.
  Space commerce / John L. McLucas.
    p. cm. — (Frontiers of space)
  Includes bibliographical references and index.
  ISBN 0-674-83020-2 (alk. paper)
  1. Space industrialization—United States. 2. Artificial
  satellites in telecommunication—United States. I. Title.
  II. Series.
HD9711.75.U62M36 1991   90-43957
338.0919—dc20   CIP

CONTENTS

FOREWORD BY ARTHUR C. CLARKE

For more than half a century, I have been writing about the promise of space—the opportunities for enriching life on earth through space technology and ultimately through space exploration. During that time I have been lucky enough to see many of my predictions come true, often more rapidly and more fully than I had dared to hope. It is therefore with great pleasure that I welcome this book on space commerce, written by Dr. John McLucas, whose own career in aerospace and in satellite communications has made an outstanding contribution not only to his country but to the world. In recognition of his achievements, the Centre for Modern Technologies, near my home in Sri Lanka, named John McLucas recipient of the first Arthur Clarke Award.

Communications satellites may have added to the cacophony of the airwaves, but as *Space Commerce* makes abundantly clear, they have also demonstrated their ability to improve the quality of life on earth and to make a profit for their investors. Undoubtedly Intelsat is the communications system that has provided the greatest financial returns for its 119 member countries, while serving their international communications needs along with those of another 65 affiliated countries, principalities, and islands. Some 30 of those countries also use it for various domestic services. Intersputnik serves a smaller community of Soviet-related countries.

Regional satellite communications systems are also coming into their own; witness Eutelsat, organized by the 26 member countries of the European postal and telecommunications organization, and Arabsat, which includes some 20 Arab states. Eutelsat is now blazing new political trails, as Poland becomes the first member from Eastern Europe. If the terms Eastern Europe and Western Europe gradually become obsolete, communications satellites will take a share of the credit. They provide a textbook case for the value of international cooperation by demonstrating—even to those late-twentieth-century dinosaurs called sovereign states—that global interests can coincide with self-interest.

The Inmarsat system—a London-based consortium owned by 58 member countries—operates a maritime satellite communications system serving sea-going vessels and aircraft. It has made the greatest contribution to safety (and efficiency) at sea since the invention of radio itself. And when it comes to safety, no one can be more in favor of satellites than the over one thousand people who owe their lives to the search and rescue system called Cospas/Sarsat. This system combines the support of its four founders—France, Canada, the United States, and the USSR—while its beneficiaries include all those countries that subscribe to the use of the standardized emergency beacons. Such space-based systems provide a resounding answer to those critics who say, "Why spend money in space, when there are so many problems to be solved here on earth?" Many of the problems on earth can be solved only by spending *more* money in space.

For example, some of our most pressing environmental concerns cannot be properly analyzed and remedied without data obtained from remote-sensing satellites in space. So far, we have barely begun to appreciate the value of these assets. Landsat, the first American Earth Resources Satellite, and SPOT, the French system using newer technology, are forerunners of systems bound to revolutionize many aspects of our lives and benefit whole industries: forestry, agriculture, mining, fisheries, town planning, and flood control, to name just a few.

In this book McLucas examines remote sensing and other potential space ventures from the viewpoint of the commercial developer and comments upon which kinds of activities may be made into profitable enterprises, and how that might come about. After communications satellites, McLucas thinks the launch business promises to be the second commercial success, though getting to that point may require an international agreement not now in existence. However, as McLucas clearly recognizes, focusing on commercial success imposes a very limited view on human activity; if we did only those things that showed promise of immediate cash returns, life would be dull indeed. Moreover, those timid souls who insist on a guarantee of profits right from the beginning will frequently be disappointed in space commerce; only the bold can hope to—or deserve to—benefit from new ventures in space.

One of the boldest ventures in space ever undertaken was the Apollo mission. But manned space flight to other planets intrigued us long before astronauts set foot on the moon. Today, the newspapers are full of articles in which both American and Soviet critics challenge the value of manned space activities and promote robotic space exploration as an alternative. Yet, policymakers in both countries continue to believe in the concept of putting men and women into space. If interplanetary travel is ever to be a realistic goal, much work must first be done on new technologies such as those required to recycle food and water. We cannot go on interplanetary voyages until we develop ways to live in closed-loop life-support systems. And if we are to remain healthy for years in space, we must discover why our bones and muscles atrophy and find ways of preventing this. Permanently manned habitats in orbit—perhaps developed and run for profit by private corporations—are a critical way to study these questions. If past experience is any guide, we can expect that the knowledge gained from these experiments in space will lead to technological, environmental, and medical benefits for people who will never set foot on another planet.

In an early novel, *Prelude to Space* (1947), I coined the hopeful slogan: "We will take no frontiers into Space." Ten years later, Sputnik proved that from orbit, national boundaries were meaningless—and the astronauts and cosmonauts soon confirmed that they were invisible. Since then, numerous satellites have helped to merge the human race into one global family—though not yet, alas, a totally peaceful one. Military reconnaissance satellites—the older, richer cousin of the remote-sensing satellites—may well have averted World War III by removing many of the great powers' fears of one another. A system based on the military reconnaissance satellites which I have advocated for years—called Peacesat—could extend the current tepid truce between the major powers of the East and the West to the rest of the world, by monitoring potential political catastrophes. The

United States and the USSR seem to have belatedly come to recognize the folly of challenging each other's guns, tanks, and nuclear arsenals. It is high time they adapted their space weapons—the rockets that carry nuclear warheads—to launch peaceful satellites, a task that could be accomplished with equal ease. This is the obvious way to reap dividends from the billions spent on ICBM development. The Third World, where most of today's conflicts are taking place, could benefit enormously from the technologies of peaceful satellites.

The exploitation of space for peace, pleasure, and profit holds much excitement. I hope to enjoy watching and recording such events for many years to come.

SPACE COMMERCE

INTRODUCTION

This book attempts to separate fact from fiction in a field where the two are frequently hard to distinguish. For about five years before the tragic loss of the space shuttle *Challenger,* discussions of commercial uses of space became a growth industry on the lecture circuit and in the trade press. Expectations ran high as speaker after speaker mounted the rostrum to describe the dawning of the commercial space age, during which space industries would garner billions of dollars per year from technical products and services. In the space-age pharmaceutical industry, for example, medicines would be synthesized in the microgravity environment of outer space, using an exotic technique called electrophoresis; almost magical drugs, hitherto unavailable, would quickly revolutionize medicine. In the high vacuum of space, the absence of convection currents in molten materials would permit production of substances of extremely high purity that would exhibit wholly new properties. Orbital laboratories would create new solid-state materials for advanced microcircuits, based on crystals and alloys that could not be grown or manufactured on the ground. Integrated circuits made from such new materials as gallium arsenide would be more resistant to damage by heat and radiation and could operate in all environments, including hostile parts of outer space.

At the height of this enthusiasm, dozens of new companies sprang up, and established corporations set up entrepreneurial departments to capitalize on the cornucopia of opportunities. I myself joined a prestigious board of directors of a company created by Willard R. Rockwell, Jr. (son and namesake of the founder of Rockwell International), called General Space Corporation. It was part of a conglomerate of space-related companies that Rockwell expected would soon become the space-age counterpart of General Electric, General Dynamics, and General Motors.

Astrotech Space Operations (ASO), an affiliate of General Space, set up shop at Cape Canaveral to provide pre-launch services for the huge numbers of commercial communications satellites that would soon be put into orbit. In addition to these services, General Space would supply electric power to the space shuttle from power stations in permanent orbit around the earth. At these space depots, shuttles could dock and receive not just power but oxygen, food, even Coke and Pepsi. Space mechanics would be on stand-by alert for dispatch to ailing spacecraft. Our company would also grow gallium arsenide crystals and other new materials in dedicated space factories. And because NASA's space shuttle fleet would soon be hard-pressed to ferry enough supplies into orbit to meet the demand for new products and services, General Space might even contract with NASA to supply additional shuttles. Venture capital would be raised in billion-dollar chunks to finance the space hardware necessary for such activities.

During this period of euphoria, President Reagan established the National Commission on Space to point the way to the burgeoning opportunities of the future. On behalf of the Commission (known as the Paine Commission after its chairman, Dr. Thomas Paine), several of us chaired hearings around the country to ensure that a wide variety of opinions about emerging space possibilities could be given adequate attention. The Commission's report was scheduled for release early in 1986.

Then came the *Challenger* accident—preempting the futuristic Paine report. The bubble of optimism about America's fledgling space industries burst. Criticism turned to recriminations about the tragedy—who was responsible, and how things could have gone so terribly wrong when everything seemed so right. The expected release date for the Space Commission's report came and went without event, while all eyes focused on the hearings and revelations of the Rogers Commission, which had been set up to analyze the *Challenger* accident.

It seemed that a scapegoat was needed to assuage our guilt for the deaths of the *Challenger* crew. The villain was quickly identified: it was NASA, once the dynamic embodiment of American genius, but now grown fat and stodgy in middle age. Our pride and joy for 25 years of space spectaculars, planetary exploration, moonwalks, and

spacewalks, NASA was suddenly flat on its back. The American public—seemingly always eager to praise a winner and abandon a
loser—turned away from space.

To make matters worse, several of the United States' dwindling
stockpile of expendable launchers had exploded or veered off course,
leaving us with no large booster capability. In effect, we were
grounded for the first time in many years. It was little consolation that
our major competitor for lifting payloads into space, Europe's Ariane
rocket, also suffered losses and was grounded.

Many newly formed companies went on hold. A half dozen of us
resigned from the board of General Space to spend time elsewhere.
Others buckled down to wait out the long night of America's absence
from the launch pad.

Meanwhile, the Soviets continued their impressive progression of
manned space activities. Shortly after the *Challenger* exploded, the
Soviets orbited Mir, a new and larger space station, and continued an
apparently effortless launch rate of almost 100 spacecraft per year.
Their cosmonauts set record after record for space flight, accumulating
four times as many hours of manned flight as we had. Individual cosmonauts stayed in orbit longer and longer, with some eventually passing one year.

When I and some other Americans accepted an invitation to attend
the Moscow conference commemorating the thirtieth anniversary of
the October 1957 launch of Sputnik—the first artificial satellite to be
launched in space—we found ourselves surrounded by a smiling and
confident group of cosmonauts, all basking in the new glow of *glasnost*.
Roald Sagdeyev, head of IKI, the Soviet Institute for Space Research,
and our host at the conference, saw his picture prominently displayed
in *Time* magazine (October 5, 1987) and found himself elected a member of the Supreme Soviet. (In March 1989, the Soviets reorganized
and created the Congress of People's Deputies, of which Sagdeyev is
now an elected member.) *Time* declared that the United States had
become number two in space, and suggested that other nations were
challenging that position. Even China was portrayed as more on the
move than the United States.

Finally, in February 1988, after two years of hesitation, the White
House issued a statement outlining a new space policy. It renewed our
commitment to space, calling for exploration of the solar system and
for habitation of space beyond the earth, and it envisaged a growth of

commercial activities in space. All in all, the new policy statement was encouraging in that it indicated a renewed commitment to space activities on the part of the federal government, and a renewed interest in aiding the growth of space commerce.

The new policy reiterated that a space station was to be a permanent feature of our space program. The space shuttle fleet, with a replacement for *Challenger* already on order, was to be the upgraded ferry to the space station and the key to manned space flight. As a matter of policy, NASA and other government agencies would be required to buy expendable launch vehicles from commercial suppliers, and indeed the Air Force immediately ordered a multiyear supply of the "big three" expendable launch vehicles—Titan, Atlas, and Delta—from commercial companies.

With commercial expendable rocket makers ready to supply launch services, and with space shuttles redesigned to correct most of the original flaws, we may now say with confidence that America is back in space. But as a sadder and wiser nation, let us hope that the hype of the years preceding the *Challenger* accident will not return. Instead, we must adopt a realistic attitude about space and its possibilities, one befitting the risks, potential rewards, and serious challenges that confront us.

▪ WHAT IS SPACE COMMERCE?

I define "space commerce" or "commercial space" to be those activities in which private companies put their money at risk to offer goods and services that depend on having satellites in orbit. By contrast, *Webster's New World Dictionary* defines commerce as simply "the buying and selling of goods, especially when done on a large scale between cities, states or countries; trade." If a company gets a typical government contract to build a satellite, that activity might fit Webster's definition of space commerce, but it does not fit mine. My definition rules out the ordinary government contract because the government usually reserves the right to change the specifications any time it wants; in exchange for that right, the company probably gets progress payments every month for the work done to that date. On the other hand, I consider that a company *is* participating in space commerce if it gets a contract to build a similar satellite for a private user. The user expects the satellite supplier to deliver the satellite for the quoted price. If NASA bought a copy of a satellite built for a commercial customer, the satellite supplier might consider it a commercial order if the deal was done as an off-the-shelf purchase. Hughes, General Electric, and (the recently sold) Ford Aerospace are companies that build satellites for both government and private customers, and so they are in space commerce and also in the government-contract business. The same can be said for General Dynamics, Martin Marietta, and McDonnell Douglas, all of which sell rockets to launch satellites for both government and private users.

We must try to remember that many countries view the interplay of business and government differently than does the United States. Any workable definition of space commerce must ensure that these nonmarket economies are still included, because we face them competitively every day in many fields of space commerce. I say "try" because it is a somewhat slippery process; we think we know what private enterprise is and how companies are supposed to do business in our economy. But there are some fields where the old definitions do not seem to fit the realities of today. Space commerce is one of those cases. In the chapter on launches, we will get into the differences between commercial launches by U.S. companies on one end of the spectrum and those by China, a nonmarket economy, on the other, with the European Ariane consortium somewhere in the middle of the spectrum. But let's face it: there are as many definitions of what constitutes commercial sales as there are people in the business, and space commerce is no different, as we will discover in the course of this book.

The two largest areas of activity in space commerce are (1) the construction of communications satellites and their ground-control stations and (2) the sale or lease of the communications services which these satellites make possible. The third largest is (3) launching satellites into orbit; it is a somewhat smaller industry in terms of dollar cost, but it is obviously an essential component of space business.

After these three items come projects that have yet to prove themselves as commercial ventures but which nevertheless offer intriguing opportunities. They are (4) remote sensing, which entails taking complex measurements of the earth's atmosphere, land surfaces, and oceans in order to discern weather patterns, climatic changes, crop yields, destruction of the rain forest, expansion of desert areas, increase in carbon dioxide and other so-called greenhouse gases, deterioration of the ozone shield, and a host of other natural and man-made changes on our planet; (5) precise navigation by satellites, serving ships, boats, airplanes, trucks, even automobiles; (6) designing, constructing, and launching space habitats and laboratories for scientific research and product development, and maintaining their human operators and robots; (7) conducting life-science experiments on the effects that space habitation has on humans; (8) processing materials in the high-vacuum and microgravity environment of space; (9) servicing and repairing payloads in space.

In addition to these examples of commercial activities in space, there are a number of quasi-space commerce activities that support government-owned space assets. These are the space equivalent of what is called "contracting out" on the ground. As a matter of policy, the federal government must contract out certain types of services—for example, housekeeping, food services, and grass-cutting at government facilities—when a private contractor is willing to do the work for much less than the government's cost. In space, as on the ground, there are a number of activities that could be contracted out to private businesses. The list includes launching government payloads on commercial rockets; providing power and critical supplies to the space shuttle; launching payloads from the shuttle to higher orbit; providing a man-tended laboratory for government use, serviced by the shuttle; converting the shuttle's external tank for use as a laboratory, warehouse, or even trash can; repairing and refuelling government-owned spacecraft.

The combination of activities on the commercial list and those on the contracting-out list adds up to some exciting business opportunities in space. But whether these prospects become realities depends on many factors: the imagination of the entrepreneurial community, the availability of risk capital, and the willingness of the federal government to create a political and economic environment conducive to success in space commerce. For years, the U.S. government was unique in attempting to foster a commercial space program. With few exceptions, there are no space programs in other countries that are not owned and operated by the government, or have major government participation. Unfortunately, creating the climate for success in space commerce has not been a high priority of U.S. policy except for the communications satellite industry—a success story for the last 25 years. U.S. space policy with regard to other opportunities in space has not been well articulated in its details, and support for commercial space activities has been sporadic and frequently confused. Because of the great cost of getting into space and operating there, governmental policy still determines what happens in space to an overwhelming degree. In our country, lack of clear presidential leadership—and, more important, lack of the necessary underpinning of support—can be fatal.

To take but one example, private businesses in the United States that specialize in launching satellites cannot succeed without some protection from foreign competition; they cannot compete head-on against nonmarket economies such as China and the Soviet Union, each of which can set prices at any level they like—and neither of which has a means of relating price to cost in the manner that market economies do. These nations' need for hard currency and their desire for acceptance as space powers will usually overwhelm any commercial realism that would otherwise control prices. And in other commercial areas such as remote sensing where we are now pitted against Europe—soon to be joined by Japan and others—the policy of the U.S. government must ensure a level playing field and the development of new technologies if American companies are to be able to expand their operations into space with any reasonable degree of success.

How, then, does the U.S. commercial space industry fare today? In the following chapters, I will first describe the communications satellite industry; then I will look at the major commercial space activities listed above and venture some guesses about their chances for development. Some seem to have much promise; others are very difficult to assess. I will also review the crucial role that the Department of Defense (DOD) and the civil space agencies—the National Aeronautics and Space Administration (NASA) and the National Oceanic and Atmospheric Administration (NOAA)—play in the success of space commerce.

It is impossible to understand the potential and risks of doing business in space without appreciating the political environment in which these industries must operate. Because of the strategic importance of space, government policy in regard to uses of space is a critical factor in the success of any commercial ventures there. In many ways, our government's space policy has been poorly conceived and at times counterproductive. In tracing the course of our space program I will not resist making a few suggestions that might put our nation on a better track in space.

■ THE CIVIL AGENCIES AND THE MILITARY

Commercial space is part of a triad of space activities, the other two being military space and civil space. The military is the largest, with a budget of about $25 billion per year. The civil program—mainly NASA's but also including NOAA's—amounts to about $15 billion. Commercial space, at about $6 billion, is the child of these behemoths. It never could have existed if they had not come first and paved the way with their huge investments in developing materials, structures, rocket engines, spacecraft, sensors, control systems, computers, launch facilities, and tracking stations—all of which are essential to space commerce but none of which could have been financially justified by the possibilities of space commerce alone.

The Space Age began on October 4, 1957, when the Soviet Union put the world's first artificial satellite in orbit. Sputnik was a coup of major importance. In the West, the conventional wisdom held that the Soviets were locked in a backward country—they had used horse-drawn artillery as recently as World War II, for example. We in the United States were sure that such a nation was incapable of any major technological achievement, especially a feat as mind-boggling as opening the frontier of space.

At first, the United States was preoccupied with the possible military significance of what the Soviets had done. Would space be used as a platform from which to attack us, or perhaps as a place to store weapons? Some Soviet actions did indeed lead in this direction, but it soon became apparent that the most important strategic use of space was as a place from which to view activities around the globe and gain invaluable intelligence.

The earliest "spy satellites" were regarded by the public as sinister, if not evil, devices. Yet it was the availability of these satellites that permitted President Eisenhower in 1960 to pledge to the Soviets that we would never again fly U-2s over their territory, as Gary Powers and others had done. Beginning in the 1970s, our satellites took on a new aura of peacekeeping when it became generally known that their capabilities allowed us to enter into arms control agreements that we would otherwise have found unverifiable and therefore unacceptable. The pejorative term "spy satellites" gave way to "national technical means of verification," and those of us involved in space activities no longer felt we had to apologize for the way we made our living.

Military satellites enable us to follow the development and testing of new weapons and the construction of new military bases, and thereby monitor the balance of power. We can detect movements of troops, tanks, ships, and airplanes and be on the alert for possible actions against us or our friends anywhere in the world. Our "eyes in the sky" can even peer inside many government offices and glean information about the intentions of other nations. An important class of satellites is devoted solely to warning of launches of ballistic missiles from land or sea. By means of its so-called space assets, our government can stay in continuous contact with our own military forces even when they are on the move in some remote location, and can deploy them as necessary.

Because of the overriding strategic importance of military intelligence, space is not just another place to do business. Business people may decide that there is money to be made in space, but they cannot proceed with their plans unless key national security officials approve them. While our government seeks to encourage space commerce, it recognizes the need to police carefully what goes on there. It has an international obligation to supervise U.S. space activities under the Outer Space Treaty of 1967. Launching a satellite requires a rocket very similar to an intercontinental ballistic missile (ICBM); our government must assure other countries that the rocket headed their way carries nothing threatening. Any private entity wanting to launch rockets that will fly over other countries must first get a license from the Department of Transportation and must register the flight with the International Frequency Registration Board, an agency of the United Nations. Communications satellites are treated as munitions by the Office of Munitions Control (recently renamed Office of Defense Trade Control); their sale to foreign countries, or export for launch by foreign countries, must be specifically approved by the Defense, State, and Commerce departments, and Congress reserves the right to participate in such decisions.

The United States' initial interest in space was played out against the backdrop of the Cold War missile race. Both we and the Soviets mounted major efforts to develop military rockets—the ICBMs which by their awesome destructive power have kept an uneasy peace between our two nations for 40 years. But the United States soon realized that space had many dimensions, and that much good science could be done there. Recognizing the need for balance between military and civil space efforts, and the fact that space held great promise for improving life on earth, President Eisenhower and the Congress joined together in 1958 to create NASA as our civil space agency. The Space Act of that year stated that NASA's task was to explore the universe and find out as much as possible about how Earth functions in the vast sea of space. The act called for NASA to make its results known to all. By contrast, the Soviet exploration of space was entrusted solely to the military. It took the Soviets almost 30 years to create their civil space agency, Glavkosmos; its degree of independence from the military is not yet fully understood in the West.

NASA has brought us close-up views of many wonders of our solar system, including the surfaces of Mercury and Mars, the giant red spot on Jupiter, the marvelous rings of Saturn. Taking advantage of the only opportunity in 175 years to have a single spacecraft visit almost all the outer planets, NASA developed the *Voyagers* and sent them on the Grand Tour. We have now received high-quality data from observations of Neptune, almost 3 billion miles away. Pluto is the only planet not yet visited by a U.S. spacecraft.

NASA also gave us impressive television pictures of the weather, showed us the extent of chlorophyll in the oceans, the state of maturity of crops worldwide, the level of water in lakes, the sprawl of the world's cities, the snow buildup on mountains, and the subsequent flooding from the spring thaw. Although several of these early weather- and surface-monitoring programs have now been turned over to NOAA, it was NASA that first made them possible.

But clearly the most visible and inspiring of NASA's remarkable accomplishments in space was the Apollo program, which took 24 people to the moon and back. Those flights represented an engineering feat that even today remains singularly impressive. Putting human beings on another celestial body was indeed a spectacular achievement, but equally remarkable was the effect that seeing the beautiful blue planet Earth from the perspective of the moon's desolate landscape has had on our thinking about Spaceship Earth, the only habitat in the solar system known to be friendly to living things.

Many people think that NASA has been on a starvation diet during the last decade, but in fact several important programs have been under way that will soon bear rich fruit. NASA has been preparing almost a dozen spacecraft that will help unravel many of the secrets of the solar system and the cosmos beyond. The *Magellan* mission launched in April 1989 is bringing us new information about the planet Venus, that fog-shrouded hothouse that no telescope can penetrate. *Magellan's* radar will map the planet from pole to pole, with a quality greatly superior to the selective sampling the Russians did in the mid-eighties. The *Galileo* mission to Jupiter, launched in late 1989, will explore that planet in great detail, sending probes into its gaseous surface. The Hubble Space Telescope, launched in early 1990 and flying high above earth's obscuring atmosphere, is bringing us views of the universe nearly as far back as the big bang 10 to 15 billion years ago. The 1990 launch of the spacecraft *Ulysses* on an inward journey to the sun will give us views of its poles and clues to the gigantic heartthrobs of that solar furnace—an inferno whose slightest change reverberates throughout the solar system. In addition to the space telescope, NASA plans to launch the Advanced X-ray Astronomical Facility, the Space Infrared Telescope Facility, and the Gamma Ray Observatory.

What we have done in exploring space and studying the earth from orbit has been good for mankind generally, but to understand the constraints imposed on commercial and scientific uses of space we should keep in mind that opening up space was more than a scientific enterprise—it was a contest between the Russians and ourselves. Racing the Soviets into space has driven much of what we have done there, and the competition has been carried out in both the military and civil spheres. Typically—once we got away from the Sputnik era—Soviet military spacecraft have lagged ours by 5 to 10 years in terms of comparable capabilities. This is essentially true for civil space activities as well, though the Soviets have carried out certain civil programs ahead of us, albeit with cruder spacecraft. Photographing the back side of the moon and taking radar pictures of Venus are two cases in point.

In the sphere of space commerce, by contrast, our competition thus far has come mainly from the Europeans, and belatedly the Japanese and the Chinese. But it may be premature to write off the Russians as competitors in space commerce. U.S. policy permits other countries to vie with us in the marketplace for economic and political reasons that may or may not be related to space. Who knows what policy we will adopt with respect to Soviet competition for launch business, for example? A White House announcement in August 1990 indicated that we will approve Soviet launches of some of our satellites from Cape York, Australia, provided a United States–Soviet agreement can be reached meeting the terms of the new U.S. space launch policy (still in draft form).

We simply cannot afford to forget the strategic importance of space. Nor should we forget the key role of military and civil space in our nation's economic development; much space-related activity is truly hi-tech and has exerted great leverage in driving advances throughout the economy. By contrast, many people would say that commercial space ventures have not been of sufficient size to have any impact on the nation's economic health. We will examine the potential role of space commerce in the U.S. economy in the next section.

■ THE ECONOMIC AND SOCIAL IMPACT OF SPACE COMMERCE

The manufacture of communications satellites and associated ground terminals in the United States totals about $6 billion per year. Of that figure, almost $2 billion is spent on communications satellites owned and operated by private companies. Launching them constitutes a business of about $1 billion per year. Once in orbit, satellites generate revenue in the communications industry at a level of about $3 billion a year. So, adding these together, we have a space-based telecommunications business of about $6 billion a year. This does not qualify as a large factor in our economic life.

But if we take a broader view of the economic impact of space commerce, a different picture emerges. To take an example, suppose a bar is doing a reasonable amount of business in a remote area but suffers from poor television reception. The owner buys a small-sized receive-only antenna which he places on his own property—hence the term "backyard antenna"—and with it brings in 100 channels of TV for his customers, all relayed by satellite. Business begins to boom. His growing business is not space business, but the growth is due to the existence of the space business. Another larger and more apt example is given by the cable TV industry. Unquestionably, a large fraction of cable revenues depends directly on the TV channels beamed down by satellite. Whereas the bar is selling drinks, not TV, the cable company is selling TV directly. If I pay $30 per month for cable TV service and there are 50 million homes like mine receiving TV service from satellites, a little arithmetic shows that this adds up to real money changing hands in the economy; total cable revenues are about $15 billion per year.

If we split the total amount of money involved in satellite communications between what is spent on the satellites themselves and what is spent on the ground so that those satellites can be used, we get some interesting figures. Let's consider the case of a satellite that costs $100 million by the time it gets to orbit and carries 12 to 24 programs, including a TV program that is being distributed to NBC affiliate stations nationwide—about 160 stations. The ground equipment at the NBC affiliates that enables their viewers to watch the show brought by satellite costs about one half billion dollars. Also watching the same satellite program are 7,000 cable TV systems all over the country, each with special equipment to pick up the satellite signals. This is worth about $3 billion. Also watching the same satellite are 3 million backyard antennas which cost $6 billion. So in order to take advantage of the $100 million investment in the satellite, the people on the ground have invested about $10 billion.

Spending all that money also allows the users to watch more than one satellite. Programs on about 25 satellites are regularly watched by many of these same people. Those 25 satellites have cost the industry about $2 billion, while the ground equipment used to tie in to the satellites is perhaps five times as large. Again, we see a large multiplier between what is spent in orbit and what is spent on the ground. (Some people argue that all the money is spent on the ground, and none is spent in space, but I will let that argument pass.)

However we may measure the economic impact of space commerce, the numbers will not do justice to the importance of the communications industry to our well-being. Satellites have come to be a key factor in allowing us to maintain contact with friend and foe alike around the globe. Worldwide communication, made possible by the opening up of space, is one of the most significant sociological developments of the twentieth century. Anything having to do with communications has an impact not measured in dollar terms alone. The reason we have a post office service is that our economy is dependent on reliable mail delivery. We should not measure the value of the post office by the size of its revenue; in fact, one can make the case that its value depends on keeping its costs to a minimum. So the smaller it is in dollar terms, the better off we are. Our morning newspaper may cost only 25 cents, but does that tell us anything about what life would be like without newspapers?

Aside from the communications industry, a number of lesser commercial space activities have thus far had little direct economic impact, but the potential is there for a large indirect effect. Remote sensing, for example, consists of flying satellites to gather data from space. Two kinds of satellites are now in use. One kind gathers data about the weather and the atmosphere. This group includes TIROS, Nimbus, and GOES satellites. The second gathers data about the land surface. These are the Landsat satellites. So far, such systems are not commercial; they are owned and operated by the government or by a contractor operating under a government contract. It costs the government something like $100 million per year to operate its civil weather satellites and another $100 million to keep its Landsat program going. Yet knowing more about the weather can have economic and social importance out of all proportion to what it costs to collect weather data in space. Better warning of storms can save billions of dollars in property loss and personal injury.

In parallel with the operation of Landsat, there is a private "cottage industry" or value-added industry engaged in creating various products by processing the data collected by the Landsat satellites. This industry employs thousands of people, and the revenue has been estimated to be between $100 and $200 million per year. Practitioners who had only Landsat data to work with in the past now may use data from SPOT, the satellite orbited by France in 1986, and SPOT II, which was launched early in 1990.

Several other categories of space industry would have to be included if we wanted to get the total value of space business. But since all other space-related activities generate revenues in the range of a few to 25 million dollars each, they do not change the totals very much. Even so, I am sure that one or two of these industries will grow enough in the next few years to achieve some economic significance. For example, navigation/location services show signs of being ripe for rapid growth.

Usually space activities of all kinds are out of sight and out of mind.

We have come to take space for granted, now that we have passed the thirtieth anniversary of the beginning of the Space Age. But from time to time, space spectaculars and accidents take center stage—as when the space shuttle *Challenger* crashed, or when a Soviet cosmonaut became the first human being to live a whole year in orbit outside the atmosphere. Like space in general, space commerce also suffers from being invisible most of the time. Communications satellites work so well that very few people even know about the occasional problems when a satellite refuses to point at the earth, as it must do to be functional. A primary purpose of this book is to give an account of just how useful these silent enterprises are and how the existing successful businesses may be joined by other promising ventures in space.

1

THE EARLY YEARS OF SPACE-BASED COMMUNICATION

Satellite communications is the largest and most visible activity in space. It is also the one space industry that has established itself as a business success, with an assured and growing future. In this and the following two chapters I will concentrate on satellite communications in some detail, because our history of achievements in this field has much to teach us about the potential of other commercial ventures in space.

The story began in 1945, when Arthur C. Clarke, an officer in the Royal Air Force and later the author of such popular works of science fiction as *2001, A Space Odyssey,* wrote a paper entitled "Extraterrestrial Relays," which described an idea for a worldwide communications system. According to his scheme, three satellites positioned equidistant from one another over the equator at an altitude of 22,300 miles and linked by radio to one another and to the ground would allow anyone on earth to reach anyone else on earth—no matter how far away—by tying into this radio network. The particular orbital path that Clarke described is a geosynchronous orbit, which means that a satellite at that specific altitude above the equator orbits the earth at precisely the same speed as the earth rotates. Therefore the satellite stays above the same spot on the earth at all times, allowing radio signals to be relayed through it without interruption.

Clarke's paper attracted little attention at the time. The satellites he described—which he called space stations—were impractical for several reasons. To keep the transmitters in the spacecraft operating, Clarke thought that several repairmen would have to be on board with a large supply of vacuum tubes, power supplies, and tools, not to mention a complete habitat to support them for several years at a time. For many reasons, it seemed that the system he described could not be put into space any time this century—a situation he did not fail to point out. Little did Clarke, or any of the rest of us, know that just three years later Bell Labs would announce the invention of the transistor, and many hitherto impractical ideas like unattended, long-endurance satellites would become possible.

Clarke's paper was published just at the end of World War II, when people were dreaming new dreams about what life could be like if we directed all the ingenuity that had been devoted to fighting the war toward more peaceful purposes. The combination of pent-up demands within the civilian economy and advancements in technology that had come out of the war suggested new opportunities to those in communications, which was Clarke's field. He was in fact the chief technical officer at the British field tests on the first Ground Controlled Approach Radar, the development of which he shared with the Americans Luis Alvarez, George Comstock, and Bert Fowler. Once the war was over, people in the communications business got interested in improving the existing communications system and began looking at all reasonable possibilities, including the use of satellites.

Up until that time, ordinary people were rather happy with the communications system we already had. We could talk by telephone to our friends within a few miles using local service, or we could call across the continent using long distance. The telephone company easily took care of our needs with their system of wires, cables, electronic amplifiers, and so on. As more people began calling long distance, AT&T put microwave towers on the hilltops, establishing radio links of very large capacity to accommodate the growing traffic. By the end of the war, transcontinental telephone links had been in place for decades, so Clarke's ideas on satellites did not seem to have much significance for our national needs. The only satellite most people thought much about was the moon, and even then they didn't call it by any fancy name like satellite.

Calling overseas was a different matter altogether. Guglielmo Marconi had established the first radio links between Europe and America in 1915, but despite many improvements, communication by radio was usually unreliable, noisy, and expensive. Service was so poor that very few people bothered to try calling overseas unless it became a matter of life and death. So a link from the United States to Europe was the first application where the possibility of using satellites was seriously considered.

Another way of talking to Europe was also being looked at: the undersea cable. Wires strung across the countryside had given us very satisfactory long-distance service. Many people in Europe also bene-fited from a fairly good wire-based long-distance service, even though we Americans thought they could do a lot better. It was reasonable to think that, with good engineering, we ought to be able to build a voice-quality cable of adequate reliability and lay it under the ocean. After all, a telegraph cable under the Atlantic had provided service to Europe for nearly a century, and a telephone cable under the English Channel further demonstrated the feasibility of this approach by providing tele-phone service between the British Isles and Europe. So, ten years after the war, we laid the first transatlantic voice cable to link the United States and Scotland.

The transatlantic telephone cable, TAT-1, which went into use in 1956, gave us reliable voice connections to Europe for the first time. The one major drawback was its limited capacity. It was designed to handle 36 telephone calls at any one time—and no television at all. As the traffic grew, the question that quickly arose was how to handle the increasing volume. Should we lay more undersea cables, or should we instead consider putting up satellites? If we were to take satellites seriously, we would have to satisfy ourselves that we knew how to build them and also how to put them into orbit. There was considerable doubt in the minds of many reputable scientists and engineers that we could do either of these things.

▪ FROM GUIDED MISSILES TO SPUTNIK

During the first decade following the war, the military had given some thought to building and launching satellites, but its main preoccupation was with guided missiles. The possibility of replacing inaccurate bombs and rockets with precision devices that would go where they were meant to go was an intriguing one, and so the military was busy designing guided missiles of various sizes and ranges. Any effort de-voted to satellites as such was quite small by comparison. Even though the RAND Corporation had written a report on a globe-circling space-ship as early as 1946, few people paid attention to such an impractical idea.

One problem was the term "spaceship," which connoted a large vehicle that would demand a massive rocket to launch it. To take hundreds of tons into an orbit thousands of miles above the earth was far beyond the range of the technology then available. Of course we were all too familiar with the V-2 rockets the Germans had fired at London; rockets of this size could take a few hundred pounds a few hundred miles, but they could never launch a spaceship. A team of German scientists and engineers, working at a small place called Peenemunde, on the Baltic coast in eastern Germany, had developed the V-2, based on the work of a German scientist named Hermann Oberth. Oberth, in turn, had drawn on the work of a little-known Russian scientist, Konstantin Tsiolkovsky, who had laid a solid foundation for rocketry near the turn of the century through his pioneering theoretical work. Oberth also knew the visionary work of Robert Goddard, an American who established a scientific basis for rocketry through both his theoretical and experimental work in the twenties.

When the Peenemunde team was overrun by Russian troops at the end of the war, the majority of its members, including Wernher von Braun, were captured by the Americans; others ended up in Russian hands. Both countries began to capitalize on the knowledge of these scientists to augment rocket work they were already doing. These developments in guided missiles by the American and Soviet military were not unknown to the people of the United States, but only those who were watching events carefully would ever suspect that this work would eventually lead to satellite launches.

Could a wartime V-2 rocket pave the way for launching a satellite? We got an answer of sorts to that question on October 4, 1957. The Russians shocked the world that day with their launch of Sputnik.

Geologists and other scientists knew that both the Russians and the Americans had plans to put up satellites under a program called the International Geophysical Year. The IGY was the period 1957–58 when scientists in many countries intended to measure dozens of the earth's properties, in the atmosphere and oceans as well as over land masses. Many scientists but few other citizens had thought about how well we could see and analyze the earth from the bird's-eye view that a satellite would enjoy. Many patterns that would be totally invisible from the ground could be easily observed from space.

The IGY was conducted at a painstaking pace appropriate for science. But this tortoise-like speed seemed totally out of place when we awoke to find a Soviet satellite spinning over our heads. Those of us involved in work with satellites endured many sleepless nights as the American satellite program inched its way toward success. Once the Russians had done it, why were we dragging our feet? Bureaucratic answers did not quell our deep frustration.

Finally, a small American satellite made it into orbit. From today's perspective, the fact that the Russians beat us into space by four months does not seem so significant, but at the time that four-month delay seemed interminable. That first Explorer went up in late January 1958, and we breathed a belated sigh of relief. Yet the Russian leap into space had damaged our psyches so badly that the shock did not go away fully until Neil Armstrong set foot on the moon 10 years later. It wasn't just Sputnik; the early years of the Space Age saw us suffer one humiliation after another as the Russians put the first dog, the first man, and the first woman in space.

▪ THE FIRST COMMUNICATIONS SATELLITES

The early satellites placed in orbit by both the Soviets and the Americans were not communications satellites. Sputnik carried a small transmitter that allowed it to do minor propagation tests, but its real mission was to show that the Soviets could reach space and operate a payload there. In a few days, Sputnik's transmitter was dead, its purpose accomplished.

The IGY satellites such as Explorer were put up to collect scientific data, and they did an excellent job, including making the momentous discovery of the Van Allen belts that surround the earth. These are two zones of high-intensity radiation trapped in the earth's magnetic field, beginning at an altitude of about 500 miles and extending out thousands of miles into space. But these scientific Explorer satellites had little to do with solving our communications problems.

Communications satellites are designed to repeat on another frequency any signal that is sent up to them. Their action is similar to that of a terrestrial microwave tower which receives signals from a distant tower and repeats and relays them to the next tower in the chain. Similarly, a communications satellite just repeats and relays back to earth whatever signals it receives. Most visitors and residents in New York have noticed the TV transmitting towers on top of the Empire State Building. The towers are up there so that the signals transmitted will travel as far outside New York City as possible. The Empire State Building is 1,200 feet high. If one were to put that same TV station on an airplane 12,000 feet high, the area where its signal could be received would be 10 times greater. If one were to put the same transmitter on a satellite at 22,300 miles altitude, which is what Arthur Clarke suggested, the signal could be picked up over about one third of the earth's surface.

The main difference between receiving a TV signal from a satellite and tuning in the New York station is that the satellite signal is much weaker. There are two reasons why: the TV station transmits thousands of watts of power, whereas the satellite transmits only a few watts. Also, the satellite is several hundred times as far away from the typical receiver as the Empire State Building is, so its weak signal is diluted even further by being spread over such a large fraction of the earth's surface. A much more sensitive (and expensive) receiver is needed to pick up the weak signal from the satellite.

When he was writing his first paper in 1945, Arthur Clarke did some calculations to see whether the amount of power one would need to transmit from the satellite was within the realm of reason. The answer was yes; he figured a transmitter of about 50 watts or so could do the job. He also calculated the necessary sensitivity of the receiver on the satellite, the directivity of its antennas, and so on. These answers were all available from the radar technology that grew out of World War II—a technology to which Clarke himself had contributed.

The first communications satellites launched in this country were produced for the military. SCORE (Satellite Communications Repeater) was developed by RCA and launched piggyback on an Atlas rocket in December 1958 into a low (nongeosynchronous) orbit. It contained a receiver and tape recorder to receive and store a message from the ground during one part of its orbit; it then relayed the taped message via the satellite's transmitter to a ground station below another part of the orbit, far removed from the point of origin. There was also an immediate retransmission capability, but the store-and-repeat mode was the most useful for military purposes, in that if the orbit had been properly chosen in the beginning, the satellite could relay messages between remote points in a few hours. This system would have been useful, for example, to carry orders from the Pentagon to a commander in some remote region, and to carry intelligence in the opposite direction. SCORE transmitted a recorded Christmas greeting from President Eisenhower to the world and lasted two weeks before its batteries died.

Courier was meant to be a much more refined system than SCORE, but it was designed to accomplish the same purpose—to store and transmit later. Courier was launched in October 1960 on a Thor-Able-Star rocket. It contained solar cells so as not to depend on batteries alone, but other problems beset it. The first satellite failed to reach orbit, and the second had component failures which caused it to live only two weeks.

On the civil side, one of the pioneers in communications satellites who eventually helped to turn Arthur Clarke's idea of a worldwide communications system into a reality was Dr. John R. Pierce, Director of AT&T's Bell Telephone Laboratories (BTL). For years, he and his colleagues had worked on traveling wave tubes, a kind of transmitting tube useful for terrestrial telephones, which when satellites came along would make their transmitters very efficient. He directed Bell Labs' research on communications, including satellites, and was an important player in the days when many key decisions were being made that would affect AT&T's future.

In about 1954 Pierce had begun studying satellites and decided that they probably could play a significant role in communications. His studies persuaded him that they should not be higher than two or three thousand miles, however, because if they were higher the transmission time delay would cause an echo in conversations. In simulations where an artificial delay was inserted, Pierce found that the combination of the delay and the poor quality of echo-suppression techniques then in use lowered voice quality far below the standard people were accustomed to.

The first two satellites that Pierce experimented with were put in medium-altitude orbits of about a thousand miles high. The first was a satellite named Echo. NASA launched it in August 1960 to study atmospheric density, the pressure of the sun on a large balloon in orbit, and various communications questions posed by Bell Labs. Echo was a 100-foot-diameter mylar sphere coated with aluminum to make it reflective. At its thousand-mile altitude and 47-degree inclination, it could be seen visually over much of the world. In the early sixties, one could go out after dusk and search the night sky for this bright and unusual moving object.

Ground stations on opposite sides of the country and even across

Figure 1. Echo 1, a 100-foot-diameter gas-inflated, aluminum-coated mylar balloon, was launched into low orbit by a Delta rocket on August 12, 1960. This passive system reflected back to Earth only a tiny proportion of the energy transmitted to its mirror-like surface. Echo 1 managed to keep its general spherical shape for several years, despite bombardment by small meteoroids that heavily wrinkled its skin. (Photograph courtesy of NASA.)

the ocean were able to talk with each other by reflecting their signals off Echo's surface. But for all its promise, Echo was a limited success. Since it was merely a mirror, it reflected back to Earth a very tiny fraction of the energy striking it. A powerful transmitter and an extremely sensitive receiver were needed to make the system work. Pierce commented at the time that Echo reflected back to the receiving antenna only a millionth of a millionth of a millionth of the energy sent up to it. So there was plenty of room for improving on this passive system, and there were a number of teams with ideas on how to do it.

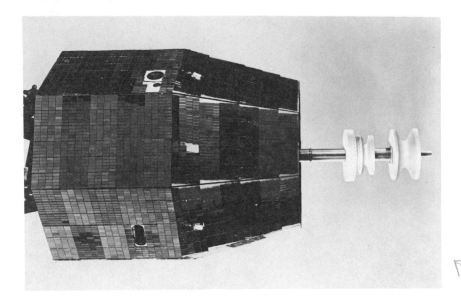

Figure 2. Telstar, the Bell System's experimental active communications satellite, served as a microwave relay in space to transmit tests of telephone conversations, data, and television programs across oceans. The satellite was covered with 3,600 solar cells. The helical antenna at the top transmitted a beacon signal for tracking by ground stations and information about the condition of the satellite for scientific purposes. It also received commands to turn on or off the satellite transmission circuits. Circling the center of the satellite were two broadband antennas, the upper one receiving at 6,390 megahertz, the lower transmitting at 4,170 megahertz. (Photograph courtesy of NASA.)

Also in 1960, John Pierce's team had conceived a design they called Telstar, an active satellite that would not suffer the power limitations of Echo. At the same time, a NASA team under Leonard Jaffe had decided to solicit bids on an active satellite, with the same objective of not being so limited in power. They received several bids, including one from Bell Labs, but instead chose RCA to construct a satellite called Relay. In addition to underwriting RCA's Relay satellite, NASA also agreed to launch Bell Labs' Telstar.

In 1962 both satellites were put into orbits a few thousand miles high, and both did what their inventors said they would do. The two were confused in the public mind, but the name Telstar captured the

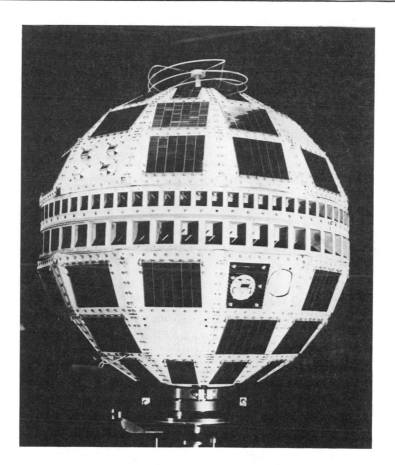

Figure 3. Relay, NASA's experimental active communications satellite, was designed to transmit television, two-way telephone, and high-speed data over transoceanic distances. The surface of the 169-pound spacecraft was covered with 8,215 solar cells that supplied power to nickel-cadmium batteries. Extending from the narrow end of Relay was a 19-inch-long wideband communications antenna. (Photograph courtesy of NASA.)

public imagination and became better known, even though Relay had a more sophisticated design and achieved a much longer life. After Telstar carried live television across the Atlantic Ocean, a survey showed that two-thirds of the population of Britain recognized the name Telstar, and over half the population had actually seen the TV broadcast (or said they had).

■ REVIVING THE CLARKE ORBIT

None of these early experimental communications satellites led directly to operational systems, but they did provide excellent data on various factors such as choice of frequencies, propagation, and stabilization and control of satellite orientation. Unfortunately, both Telstar and Relay shared a common failing. A medium-altitude satellite, because its speed of rotation is faster than the rotation of the earth, moves eastward across the sky at a rate of about 4 miles per second. So it does not take long to move out of the proper position where it can communicate simultaneously with both sides of the Atlantic. Telstar and Relay had solved the power problem by transmitting millions of times more power than Echo, but their inventors had no ready answer to the second problem. Each satellite darted quickly across the sky and disappeared over the horizon to the east a few minutes after rising in the west. About the time you got used to the high-quality signal, the satellite was gone again. It would be about two hours before it reappeared—only to go dashing off once more.

John Pierce and many others thought the way to solve the problem was to put a ring of satellites in the sky: as one disappeared to the east, another would rise in the west. About a dozen should be enough to permit continuous connections between the ground stations—provided you could keep the satellites equally spaced around the earth as they did their celestial dance. Keeping them properly spaced would not be easy. If they drifted randomly, then about two dozen would be needed to have essentially continuous contact. Each ground station would have to have two antennas, one to track the satellite that was active at the moment and another to track the satellite that would soon rise in the west. In fact, three antennas would be needed at each station so that one could be taken off the air for maintenance and repairs from time to time.

Pierce's plan was conceptually very similar to one developed by RCA a few years earlier. RCA had proposed a series of airplanes that would follow one another from New York to Paris, spaced about an hour apart. Each airplane would relay a signal to the airplane ahead, all the way across the Atlantic. About two dozen airplanes would be needed to maintain continuous contact across the ocean—half flying east, the others flying west to get in position to fly the next day. Strings of satellites would have one advantage over the airplanes: the satellites would not need to talk to one another—only to the ground. But whether you set up relays of airplanes or strings of satellites, either scheme would be terribly expensive and cumbersome.

All this talk about multiple airplanes and satellites led to an obvious question: Why not use the orbit that Arthur Clarke had written about some 15 years earlier? Orbiting above the equator at 22,300 miles altitude, each satellite would in effect sit still in the sky, directly above one particular spot on earth. If satellites were properly spaced, only three would be needed to serve the whole world. Furthermore, a ground station would need only one antenna, which could be pointed permanently at a fixed place in the sky.

There were two reasons why the Clarke orbit, as it is sometimes called, did not attract more attention in the engineering community. First, everyone knew about the delay that John Pierce had worried about and the bothersome echoes it caused. Second, there were no rockets large enough to put satellites into such a high orbit. The best rockets in the United States at that time were not very powerful. The Russians, on the other hand, were putting up rather large payloads without apparent problems of weight. Apologists said that the reason the Russians had developed such large-capacity rockets was that they had built very crude and heavy weapons that had to be lifted into space. (This is another way of saying they had not stolen as many atomic secrets as we thought!) Although there was some truth in that statement, it was not comforting to those of us who wanted to launch satellites to higher altitudes. What was needed was either a more powerful rocket or a smaller satellite, or both.

■ SYNCOM, THE FIRST GEOSYNCHRONOUS SATELLITE

Back in 1959 a design team at Hughes Aircraft headed by Harold Rosen and Don Williams had conceived a plan for building a small satellite which they judged would be light enough so that a Scout rocket could just barely lift it to geosynchronous orbit if it were launched from Jarvis Island on the equator. They were thinking of Scout as a way to reduce costs; Scout's weight limitations would enforce development of a minimal weight, ingenious design. Rosen remembers reading an article by John Pierce and Rudi Kompfner written in March 1959 that affected his thinking and started him on the path to designing a stationary geosynchronous satellite. It described the merits and demerits of both the geosynchronous orbit and lower orbits and concluded that geosynchronous satellites were not feasible for the near future because of rocket limitations, the weight of thrusters and other components needed for maintaining the position of such satellites, and the deficiencies of available guidance systems. The Hughes group looked at essentially the same technology but with the advantage of a year or so of developing their own ideas on how to reduce weight to a practical level. They arrived at the conclusion that the job was feasible, if just barely so.

One of Rosen's key contributions was his decision that the satellite would be smaller, lighter, and more easily stabilized if it were a spinner. By employing gyroscopic principles, Williams showed that they could eliminate several of the thrusters needed to keep the satellite in the right place in the sky and still keep it pointing correctly. Another colleague, Tom Hudspeth, designed extremely light-weight electronics for all the major functions on the satellite, achieving further reduction in weight. They decided that the Delta rocket was the minimum on which to stake their future and planned accordingly. It would be capable of putting 150 pounds into transfer orbit.

Transfer orbit is the intermediate step between low-earth orbit and geosynchronous orbit. To oversimplify a bit, imagine a 3-stage rocket with the first stage putting the satellite in a low circular orbit; the second stage then lifts the satellite into an elliptical orbit whose apogee is at geosynchronous altitude; the third stage is fired at apogee, circularizing the orbit at that altitude. With care, the Hughes design should come in at no more than 150 pounds, including the kick motor in the satellite that would be used to achieve circular orbit after the satellite left the Delta behind. Having demonstrated to their own satisfaction that they had met the most demanding challenges, they persuaded themselves that what they needed most was a sponsor who believed in their work.

Their boss was Dr. Allen Puckett, Vice President of the Engineering Division. In late 1960 he went on the campaign trail searching for a sponsor. He failed at the Pentagon, perhaps because they were already supporting development of Advent, a geosynchronous satellite. Although Keith Glennan at NASA was interested, he had been unable to get the Bureau of the Budget (BOB) to allocate even the $10 million needed to cover the cost of the medium-altitude Telstar satellite. Maurice Stans, the director of BOB (now the Office of Management and Budget, or OMB), said that Bell Labs was going to build Telstar anyway; we ought to let corporate America foot the bill for proving that satellites were the wave of the future. With BOB unwilling to fund even a low-altitude satellite, there was certainly no way to sponsor a high-altitude geosynchronous satellite. So Puckett struck out there also.

At about that time, the Hughes team persuaded Puckett that they should build a model at company expense. In 1961 they went to the Paris Air Show looking for support and mounted the model on the Eiffel Tower. Rosen still remembers hearing the skeptical opinion that the Eiffel Tower "is as high as your little bird will ever get!"

Perhaps that would have been the case had not the Kennedy team succeeded the Eisenhower administration in the White House. By March of 1961 NASA was successful in getting $10 million freed up to hold a design competition for a medium-altitude satellite, and this led to the development of RCA's Relay satellite. NASA also agreed to launch Telstar for AT&T, which was already engaged in building Telstar at its own expense. Aware of these developments, Puckett renewed his contacts with NASA and found that Dr. Robert Seamans, newly arrived at NASA and the top technical official there, had been intrigued with the Hughes plan from the time he had first heard about it. Seamans, who had been a professor at the Massachusetts Institute of Technology and a student of Stark Draper of inertial guidance fame, had been a manager of SAINT (Satellite Inspector), a satellite design project at RCA, before joining NASA in 1960.

John Rubel, deputy director of Research and Engineering in the Department of Defense, had been given a simulated demonstration of the Hughes satellite model—called Syncom—early in 1961. He immediately saw that this system could be a substitute for Advent, a program which was in serious technical difficulty and far over its budget. One of Rubel's roles was to coordinate DOD amd NASA activities. As such, he was in touch with Seamans, NASA's new top technical official.

While Rubel saw the problems with the Advent satellite, he thought the Advent ground stations were progressing reasonably well, so he proposed that DOD and NASA do a joint satellite development where DOD would supply the ground stations and NASA would develop the Hughes geosynchronous satellite. NASA was convinced that the Delta rocket could put Syncom into orbit, so the principal challenge was making Syncom reliable. Rubel and Seamans soon fashioned a way to do a joint demonstration of the project's feasibility, and in August 1961 Hughes was awarded the contract for satellite construction. It was not large (about $30 million), considering that such satellites had never been built nor launched into a high-altitude orbit before, but many hopes and dreams rested on it.

▪ APOLLO AND COMSAT

While these discussions were going on, the government made a momentous decision regarding the future of communications by satellite. Some historians have said that John Kennedy was looking for a way to put the Bay of Pigs fiasco of April 1961 behind him; others have said that he was interested in space for its own sake. Without question the Kennedy Administration was feeling the heat from Yuri Gagarin's successful orbit of the globe on April 12, 1961, and was looking for a way to show that the United States could outdo the Russians in space. NASA was known to be readying the Mercury capsule for Alan Shepard's flight, but it was going to be a suborbital hop of a few hundred miles and about 15 minutes' duration—hardly an impressive response to Gagarin's complete orbit of the earth. Kennedy's White House team asked NASA and DOD what they had to offer the President by way of a stunning achievement in space. James Webb, the new administrator of NASA, proposed that the United States put a man on the moon. Robert McNamara, new boss of the Pentagon, thought that was not startling enough and proposed sending a spacecraft to another planet. NASA was dead set against that proposal, but countered with the idea of adding a communications system to the moon package.

McNamara and Webb agreed that Rubel and Seamans should jointly prepare a complete proposal. The resulting document contained several initiatives, including the moon project, the development of the Saturn rocket, several communications systems, and an operational weather satellite to be based on TIROS, an experimental satellite first flown by NASA in 1960. It was felt that the combination of initiatives would go a long way toward making the President's point that we could compete with, and surpass, the Russians in space.

While the idea of putting up communications satellites found rather broad approval, the Kennedy announcement touched off heated debates between the supporters of two different concepts, one of which would lead to public ownership, one to private ownership, of the United States' satellite communications system.

On the side favoring public ownership were those who said that the technology had been developed at huge public expense and should not be handed over to greedy private companies. Senator Estes Kefauver made a name for himself by championing this viewpoint.

On the other side were two groups. One group felt that AT&T should be allowed to go ahead with its plans to build communications satellites—capitalizing on the knowledge gained from Echo and Telstar—and tie the new satellites into the existing telephone system. This viewpoint was strongly supported by Senator Robert Kerr. Another group thought we should involve private enterprise in creating and operating a satellite communications system, but we should not hand the task (and all that technology) to AT&T on a silver platter. Senator John Pastore was of this persuasion.

The intense Congressional debate resulted in a compromise embodied in the Communications Satellite Act of 1962, which called for the creation of a private company (later called the Communications Satellite Corporation, or Comsat) that would be closely regulated as to technical, economic, and foreign policy matters by the government. Half of Comsat's stock would be sold to the public, the other half reserved for purchase by communications companies like AT&T, ITT, RCA, and Western Union.

Comsat came into being in February 1963. Its charter called for creating a global system of satellite communications at the earliest possible time. Its incorporators (chosen by JFK) set about getting enough money to hire a staff and start building a system. Using its charter and mandate, they soon had established a credit line of $5 million.

■ GOING THE GEOSYNCHRONOUS ROUTE

A major decision facing Comsat early on was what type of satellites to buy for its system. (Provisions in the Satellite Act favoring maximum competition in the production of satellites strongly suggested that Comsat buy--not build—its own satellites.) Telstar and Relay had

Figure 4. President John F. Kennedy signed the Communications Satellite Act in 1962, leading to the creation of Comsat in 1963. Comsat's mission was to develop a worldwide system of communications satellites. (Photograph courtesy of Ankers Photographers, Inc.)

already been launched and operated successfully, and by default they were the obvious prototypes for what one might expect Comsat to buy. But they suffered from the failing of all medium-altitude satellites, which is that they move very fast across the sky. To carry out its responsibility of building a fulltime, economical, and reliable communications system that would provide the kind of quality service we had come to expect from the telephone company, Comsat investigated the "obvious" solution—to buy about 20 Relays or Telstars and try to put them in orbit.

Based on estimates of the cost of each satellite and the success rate of launches on the rockets of that time, Comsat arrived at a figure of $200 million as the best estimate of what would be needed to see them through the first few years of operation. So Comsat made plans to raise $200 million—half from the public and half from common carriers such as AT&T. The announcement that stock in the first company created to operate in space would soon hit the market generated a high level of excitement in the investment community. Soon the stock offering was oversubscribed, and Comsat got its money very quickly.

Dr. Joseph Charyk, the new president of Comsat, had been Undersecretary of the Air Force, where he had been in charge of the Air Force's space program. Therefore he was one of the few senior officials in town who had any significant experience with satellites. As he went about hiring an engineering staff for the new company, he considered whether Comsat should really buy 20 satellites and rockets or look for a simpler way. Although duly impressed with the success achieved by Telstar and Relay, he also knew about the project at Hughes Aircraft to build Syncom for NASA. If this concept proved itself, Syncom might form the basis for a complete operating system—not just an experimental one.

Even though it was too early to be sure, the people at Comsat knew that the alternative system of 20 satellites, for which they had already raised $200 million, would not be easy to achieve. Getting all those satellites into orbit could take a very long time, and keeping them in phase with one another might be impossible. Ground stations would be chasing satellites across the sky continuously, and each ground station would need three antennas with complex tracking mechanisms to assure good reliability, thus driving up costs. The whole endeavor would certainly be very expensive and unwieldy.

On the other hand, if a geosynchronous satellite would actually work in orbit, and if the Delta rockets could achieve a 50 percent reliability rate in launching, a geosynchronous communications satellite system could be in business after only a few launches. Two or three satellites would be enough to start up the enterprise. If they worked, not only would the cost be much less for the smaller number of satellites, but also the ground station facilities would be much less complex and fewer in number. Also the crews would not need the same level of skill to track satellites that would be essentially stationary in the sky.

But would it work? Comsat's officers knew that the Army had finally abandoned the development of Advent, their geosynchronous satellite, because of a large growth in satellite weight and huge cost overruns. By the time it was cancelled, the Army had spent as much money on development of Advent as Comsat had raised to build their whole communications system.

Sid Metzger, the first engineer hired by Charyk to head Comsat's engineering division, arrived in June 1963 and began building up his team. Four months before his arrival, NASA had launched the first Syncom satellite. The Delta rocket worked perfectly, putting the Syncom 1 satellite into transfer orbit. It was tested and found to be working satisfactorily. Unfortunately, when the satellite's apogee motor was fired to circularize the orbit, all signals were lost about two seconds before circular orbit would have been achieved. A search by astronomers turned up evidence that the satellite was still in orbit and had not exploded, as was first feared. Apparently a fuel tank had ruptured and knocked the satellite off the air.

On the second launch, which happened a month after Metzger arrived at Comsat, Syncom 2 achieved geosynchronous orbit, but the rocket's thrust was insufficient to remove the north-south inclination. This meant the satellite rocked slowly back and forth above the equator. All the components of the satellite itself worked as designed though no tests were planned to assess the acceptability of the time delay inherent in operating in the geosynchronous orbit. By August 1964 the Delta rocket had been upgraded and Syncom 3 actually achieved stationary orbit.

Syncom was a real success. The satellites weighed only 79 pounds in orbit. Everything possible had been left out to reduce weight. With no battery on board, Syncom was completely dependent on the solar cells on its periphery to supply power. Since the satellite was partially in earth's shadow only 6 weeks twice a year and then only around midnight, it worked most of the time, and it was so stable that it easily rode through the eclipses.

By December 1963, Comsat's engineers were already favoring the Syncom approach over satellites at lower altitude, but to cover itself Comsat asked AT&T and RCA to study the merits of a medium-altitude random orbit system, and it asked Space Technology Laboratories (STL) and ITT to study the medium-altitude phased orbit system. Hughes would study the geosynchronous system. In the phased orbit system, adequate control of the satellites would be maintained to keep them spaced uniformly in a ring around the earth; in the random system, they drifted freely and kept no particular relationship to each other.

While Comsat was leaning toward choosing geosynchronous satellites, the company was in an interesting position, since some Comsat board members had a vested interest in the satellite designs that had

Figure 5. Dr. Harold Rosen (right), associate manager of the Syncom program at Hughes Aircraft Company, and Meredith R. Eick, Syncom electronics project engineer, examine the geosynchronous communications satellite designed and built by Hughes for NASA. The spacecraft is shown fully assembled except for the solar cell panels. (Photograph courtesy of NASA.)

already flown. Not only was the lower-altitude orbit of Relay and Telstar the one favored by most people who supposedly knew the business, but Telstar was the brainchild of AT&T, which not only had three representatives on the 15-member Comsat Board but was also Comsat's largest stockholder (and two years later became its first customer).

One reason some people had said that the phone company should own stock in Comsat was to make sure that Comsat would have access to AT&T's knowhow. What AT&T "knew" was that the geosynchronous satellite was a bad risk, even if existing Delta rockets could put it into high orbit successfully. But to their credit, AT&T never tried to veto Comsat's decision to buy satellites based on the Syncom design, even though several of their key people felt Comsat was heading in the wrong direction. While questioning Comsat's rationale, they did not openly challenge it.

Finally, after much analysis and soul searching, Comsat placed a major bet on geosynchronous satellites by awarding Hughes a contract in March 1964 for two HS-303 satellites based on the Syncom design. The satellites were considered "experimental-operational," meaning that if they worked they would be called operational. Only a year later, in April of 1965, the first Comsat satellite—soon to be called Early Bird—went into orbit.

Tests were immediately conducted to see what the effects of the time delay might be on the quality of service. Several countries joined in the tests, and after monitoring thousands of telephone calls, the joint team declared that the quality of telephone conversation via geosynchronous satellite was satisfactory. True, the echo due to the delay from a satellite so far away was still troublesome, but soon various techniques were employed to mitigate the problem. Echo suppressors which combine outgoing and incoming signals to reduce the level of the echoes were already in use on certain long-distance land lines. Echo suppressors could be made better—and would have to be made better—if geosynchronous satellites were to be a real alternative to cable for long-distance service.

Intelsat, the new international communications consortium formed just four months after Comsat placed the order (see Chapter 2), quickly concurred in Comsat's decision to go geosynchronous, making that approach to worldwide satellite communications official.

John Pierce was a good loser. In 1968, five years after the first satellite went into geosynchronous orbit, and only three years after the first commercial service by any satellite, John wrote, "The success of Syncom makes one wonder why anyone was ever interested in low-altitude satellites." With the Syncom demonstration in 1963 of the feasibility of geosynchronous satellites, and the success of Comsat's Early Bird in 1965, we never heard any more about strings of satellites chasing each other across the sky. The Clarke orbit became the orbit of choice for communications.

▪ MORE RESEARCH AND DEVELOPMENT

Both Syncom and Early Bird were very limited in their communications capability. More developmental work to improve satellite performance was needed. To that end, NASA bought for test a total of six Applications Technology Satellites. Hughes supplied five of the series, ATS-1 through ATS-5, and Fairchild supplied ATS-6. Unfortunately, ATS-2 and -4 failed to reach orbit; ATS-1, -3, and -5 were successful.

ATS-1, launched in 1966, was a combination VHF (very high frequency, 100–300 MHz) and C-band (4–6 GHz) satellite stationed over the Pacific Ocean (see list of abbreviations for definitions). It carried traffic of educational, scientific, medical, and cultural interest. After a few years of experimental use, it served the routine message traffic needs of the islands of the Pacific for more than a decade. It was the only available inter-island link for many small islands with light traffic. One of its major advantages was that the ground equipment needed to tie into the VHF portion of the satellite was very inexpensive, so the less developed islands found its use a godsend. ATS-3, launched in 1967, was similar to ATS-1 and was used especially for remote medical purposes such as relaying data between oil platforms and mobile health centers back to hospitals on the mainland.

In 1985 ATS-1 drifted out of position so that it could no longer be seen in the Pacific region. But then in 1988 it drifted back into position and will be available until it drifts away again or dies.

While ATS-1 had advantages in terms of low-cost ground-access equipment, its VHF frequency band suffers the disadvantage of very limited capacity to handle simultaneous conversations. This is due, first, to its limited bandwidth and, second, to the need to space satellites very far apart in the sky to avoid having ground stations interfere with one another. The term bandwidth refers to the spread of the band of frequencies used in the satellite. A telephone conversation takes about 5 kilohertz of bandwidth, so a satellite with 50 kilohertz of bandwidth can handle at most 10 simultaneous conversations. New digital processing techniques have increased this number somewhat. VHF satellites have to be about 40 or 50 degrees apart in the sky so that ground antennas can distinguish between adjacent satellites. Almost all civil satellite service is now supplied at much higher frequencies—either C band or K band (11–30 GHz)—in order to avoid these two problems.

ATS-5, launched in 1969, used transmitters in C band and L band

(1–2 GHz) and thus had capacity to carry TV and many scientific payloads. ATS-6, a more advanced design launched in 1974, served for several years as an experimental system for acquiring data on L band, which is now used for maritime communications. It was stationed over the Pacific and carried out experiments for the American/ Russian Apollo-Soyuz link-up in 1975. This satellite also provided service to countries in the Indian Ocean region, and it was used by India in the SITE experiment—a test of their planned system for education and training in remote areas. The demonstration of its capabilities led to India's decision to buy the Indian Telecommunications Satellite (Insat), which is now a commercial success as well as an invaluable transmitter of both education and entertainment programming (see Chapter 3).

For about a decade after the successful ATS series, NASA gave up its experimental satellites. Caught in a budget squeeze in the early seventies, NASA surveyed the various candidates for elimination and concluded that if any field of space technology showed signs of success, it was certainly space communications. Satellite communications had every appearance of being able to survive without a NASA research program to back it up, whereas others would surely founder. So the experimental satellite program was dropped.

But by the early eighties, the United States seemed to be losing its technological lead over Europe and Japan in one field after another. A number of supporters of advanced communications research began plugging for a renewed commitment to such work. Congress was persuaded that we had let Europe and Japan gain too much of an advantage in advanced technology at our expense, and a new research and development program for communications satellites was begun. NASA contracted with several companies to build the Advanced Communications Technology Satellite (ACTS), which was begun in 1984 and is still under construction. It is scheduled for launch in 1992.

It was intended that ACTS would incorporate the research that no private company could afford to do on its own—meaning that certain technologies could not be brought to fruition rapidly enough to justify private money's being invested in them. It was assumed that NASA would take the longer view. Technologies such as on-board satellite switching, complex signal processors and computers, and hopping beams (beams which shift rapidly from one area of the country to another) were to be included, as well as advanced optical links.

ACTS' supporters hope that enough money has already been spent that its critics will not be able to stop its being completed and launched. Pessimists say that the danger now is that it might be launched but no money found to conduct the necessary tests. Optimists believe that private companies will want to use ACTS for tests and will be prepared to spend their own money to do so.

Working in parallel with NASA was the Pentagon, which sponsored the development of a number of satellites under the Lincoln Experimental Satellites program at the Massachusetts Institute of Technology's Lincoln Laboratory. The LES series improved operations at UHF (ultra-high frequency, 300–900 MHz), EHF (extremely high frequency, 30–60 GHz), and X band (7–8 GHz), while NASA pursued work at VHF and C band. Among other things, the LES series demonstrated antijam techniques, which were of no special interest to NASA. The later LES satellites concentrated on optical communications—a project NASA had originally incorporated in their ACTS satellite but unfortunately had to drop in a budget squeeze.

Both the ATS and LES programs have been successful enough to justify themselves. Whether a particular feature of satellite technology was first demonstrated by the military or the civilian side is perhaps not important; each program has benefited from the other, and many techniques are equally applicable to both types of users. The "despun" antennas on spinning satellites, cross links between satellites, and techniques to reduce interference between packages on the same satellite are all examples of new technology in the telecommunications industry that is now available to commercial as well as government-sponsored ventures.

Next we will turn to two very significant commercial applications of these many experiments with communications satellites—the development of an international communication system and the growth of domestic satellite services.

2

INTERNATIONAL COMMUNICATIONS SATELLITES

The Communications Satellite Corporation (Comsat) began life as a compromise solution to a controversy inside the U.S. government over who should own and operate the proposed international communications system. The Communications Satellite Act of 1962, crafted by the Kennedy Administration and Congress, provided a mechanism for the involvement of both the government and private corporations. A new company would be created whose purpose was to establish a global satellite communications system and to invite participation by other countries. U.S. common carriers such as AT&T, RCA, Western Union, and General Telephone would be allowed to buy collectively up to 50 percent of the common stock, with the rest of the stock to be sold to the public. The government would not invest in the company but would maintain close oversight of its operations, would provide launches for its satellites (through NASA), and would not be precluded from leasing capacity on the satellites.

The Federal Communications Commission, the State Department, and the White House Office of Telecommunications Policy were each to have a hand in overseeing Comsat's activities. In addition, the President of the United States would appoint 3 of Comsat's 15 members of the Board of Directors, the rest to be elected by the stockholders. The Communications Satellite Act was the first policy statement of the U.S. government to assume that space was an appropriate place for commercial activity.

■ THE CREATION OF INTELSAT

Shortly after its creation, Comsat sent a team of key people, hand in hand with State Department officials, to sell to other nations the idea that every country ought to join an international consortium (to be called Intelsat) that would own and operate a new global satellite communications system. Comsat's early travels netted 12 countries wanting to join.

As of early 1990, the Intelsat consortium had grown to include 119 countries, the newest member being Romania. In addition to providing service to its members, Intelsat keeps the welcome mat out for about 60 other countries and principalities that are serviced by the system. There are few restrictions on who may join—any country that is a member of the International Telecommunication Union (ITU) and is willing to accept the terms of the Intelsat intergovernmental agreement is welcome. Nonmembers may lease capacity from the system. There is scarcely a country in the world that does not use Intelsat services to some degree. A way was even found to deal with the two-China situation; when Taiwan was sitting in China's seat, Comsat handled mainland China's introduction into the use of Intelsat capacity; when the Peoples' Republic began sitting in China's seat, Comsat began handling Taiwan's traffic needs.

As the first member and organizer of the interim consortium that was to own and operate the Intelsat system from 1964 to 1973, Comsat began with a 60 percent ownership share. By 1973, its share had dropped to 50 percent. The permanent arrangements which took effect in 1973 called for the ownership shares to be recomputed each year, based on use of the system during the previous year. For the last decade or so, Comsat ownership has stabilized at about 25 percent. The next largest owner is Britain, with about 12 percent.

Intelsat operates essentially as a nonprofit cooperative for its member countries. In addition to owning and operating the satellite network, it collects working and investment capital from its members, based on its overall needs and their share of ownership, and derives revenues from the sale or lease of satellite capacity to its members and other users. These revenues are used to cover operating expenses, amortize investments, and pay its members a 14 percent return on their investment.

The governance of Intelsat is rather straightforward. The Agreement is signed by governments: they are called Parties to the Agreement and meet as the Assembly of Parties, each Party having one vote. The Signatories, which are designated by the governments, are usually the telecommunications organizations in each country (called PTTs for Post, Telegraph, and Telephone), but in the United States the signatory is Comsat. The signatories sign the Operating Agreement, put up the capital, and convene annually as the Meeting of Signatories. The governing body with responsibility for all major business and operational decisions is the Board of Governors, consisting of approximately 24 members representing most of the signatories. If there is no consensus on an issue, votes are taken and weighted according to each signatory's investment share. The Board in turn elects a chief executive, called the Director General, who is confirmed by the Assembly of Parties and supervises a full-time staff to carry out the day-to-day operations. The present director general is Dean Burch, former chairman of the FCC.

When the Intelsat definitive arrangements were being negotiated, there was a debate about how restrictive the covenant would be. While the United States sought to protect Intelsat from its members' setting up competing systems, some countries—especially in Europe—sought freedom to establish regional systems; in turn they agreed to coordinate any such action with Intelsat so as to avoid siphoning traffic away from the Intelsat system in sufficient amounts to cause "significant economic harm." Article XIV(d) specifies that a member must coordinate with Intelsat before initiating services that could appear competitive with those provided by the consortium.

The issue of competition seemed rather academic when Intelsat was created in 1964. After all, what member would spend the money to create a competitive system when Intelsat itself had not yet shown its viability? But more recently, as satellites have proved their worth, a number of countries and a few private entities have thought it advantageous to create systems more or less competitive with Intelsat. Later we will discuss some of these cases.

The growth of the Intelsat network has surpassed the wildest dreams of its founders. The first order was placed in 1964 by AT&T, for 60 voice-grade circuits from the United States to Europe. Today 120,000 circuits are leased from Intelsat by the various users of international service. Intelsat also has sold or leased some 25 percent of its total capacity for purely domestic use in some 30 countries.

The first Early Bird satellite has been followed by 35 additional satellites, 14 of which are still in service. One may ask what has happened to the satellites no longer in service. Satellites may fail for a number of reasons, but if they are properly designed they will fail gracefully. Typically, they suffer a gradual loss of power from their solar arrays. This can be dealt with by turning off one or more transmitters. Power amplifiers in the transmitters may fail, necessitating that they be turned off. Satellite orbits are disturbed by sun and moon, and gas is used to push them back where they belong. In 5 or 10 years' time they run out of gas, and their owners, not being able to keep them in position, must turn them off.

Dead satellites become wanderers through the sky in what appears to be random motion, but it is really reaction to perturbing forces. If left in a geosynchronous orbit, they tend to drift to a satellite graveyard in the sky almost directly over Sri Lanka. Arthur Clarke, who lives in Sri Lanka, often points out to visitors a cove where legend says a huge meteor fell, possibly creating a gravitational anomaly that would explain why satellites are attracted to this particular spot in the Clarke orbit.

Now that nations are becoming more environmentally conscious, users of the geosynchronous orbit are expected to use the last remaining gas to kick dying satellites out of the orbit so as not to cause future congestion or collision. While all orbits will presumably decay eventually, decay at 22,300 miles is so slow as to be essentially unmeasurable, and a satellite drifting at that altitude may take thousands or millions of years to fall down.

In addition to the older satellites already in service, Intelsat has committed about $2 billion to two series of five satellites each. Five are of the Intelsat 6 design; another five are a newer design called Intelsat 7. The first of the Intelsat 6 series was launched by an Ariane rocket in November 1989, with the rest of the five scheduled for 1990–91. The second Intelsat 6 became a piece of bad news in February 1990 when its Titan 3 launcher left it in a useless orbit at low altitude. NASA was asked to use the shuttle to retrieve and/or reequip the satellite for a second attempt. In the summer of 1990 NASA agreed to retrieve the satellite and reconfigure it for launch in late 1991 for

Figure 6. Intelsat 5 is a high-capacity satellite that carries 12,000 telephone circuits and 2 TV channels. It has 27 transponders, 6 communications antennas, 2 global coverage horns, 2 hemispheric/zone shaped beams, and 2 steerable spot beams. It stands 6 meters high, with a 15-meter wingspan, and weighs 1,000 kg in orbit. Fifteen of these satellites were built by Ford Aerospace and 13 were launched successfully between 1980 and 1989. (Photograph courtesy of Comsat.)

about $100 million; Intelsat agreed it was worth it. This whole question is of more than passing interest because of the 1988 U.S. government announcement that the shuttle would be used only for noncommercial launches; it is of even more interest because Intelsat chose to be self-insured for the Titan launch.

The first Intelsat 7 is scheduled for 1992, the 500th anniversary of Columbus' voyage to America and the year designated as the International Space Year by the United Nations and many countries.

In addition to having several hundred times the capacity of their forebears, today's communications satellites are designed to last many times as long. Early Bird's design life was only 18 months, though it served about 3 years; Intelsat 6 should achieve 13 years of life, and

Intelsat 7 is planned for 15 years of life. Early Bird weighed 80 pounds in orbit; Intelsat 6 weighs 5,000 pounds. Intelsat has decided not to keep making satellites larger and larger, on the theory that one should not put too many eggs in one basket. Intelsat 7 is thus somewhat smaller than Intelsat 6.

Early Bird could handle 240 telephone calls, but these had to be cut off to carry one television program. Intelsat 6 can handle 30,000 telephone calls plus 3 television channels. It reuses C-band frequencies as much as 6 times, and K-band frequencies twice; by using a sophisticated digital multiplication technique (DCME), the capacity can be raised from 30,000 to 120,000 circuits.

The somewhat smaller Intelsat 7 satellites will have a capacity of 18,000 telephone circuits—a figure which can be increased to 90,000 circuits using DCME—plus three TV channels. Intelsat has designed a new satellite for service beginning in 1992 which is called Intelsat K. It will serve the Atlantic Ocean region. Its principal feature will be very intense beams that will permit TV signals and other data to be received by very small antennas.

All these changes have allowed Intelsat to reduce its charges for service by 95 percent. The price for 20 circuits is now what was paid for one circuit on Early Bird. The cost of a call overseas would fall only 10 percent if the satellite part of the link were given away free of charge. Most of the cost of international telecommunications comes from terrestrial charges—especially in those countries that use international telephone revenues to subsidize domestic services of various kinds. Intelsat leases about 30 full-time television channels to its members. Comsat, in turn, provides 21 such channels to customers like NBC, ABC, CBS, CNN, and other networks in the United States. Many of these channels are used to bring news from overseas, and some are used to feed U.S. programs fulltime to overseas entities such as channels 7 and 9 in Australia amd NHK in Japan. Intelsat's total annual income from charges is about $650 million. The capacity it sells to its customers for $600 million per year is resold by the telephone companies in each country at a large markup. Some of the companies mark up service costs by 100 percent; others may add a markup of 1,000 percent. It is the prerogative of each member to do what it likes in its own country, and many countries use the system to help carry the costs of other services such as the post office. Although exact figures are hard to come by, it has been estimated by an Intelsat official that the cost to users for services based on being tied to the Intelsat network is of the order of $7 billion a year; yet Intelsat itself is responsible for very little of this cost.

By contrast, Comsat's earnings are carefully tracked by the Federal Communications Commission, and if they exceed 12 percent, Comsat is told to cut its rates to its customers to bring earnings into line with FCC standards.

Intelsat's capabilities are the backbone of the worldwide international telephone network; Intelsat reaches more countries than any other system. The telephone system carries much more traffic than point-to-point voice communications; all kinds of transmissions—facsimile and telex, television, data, new special purpose offerings, and much more—go through the telephone system. Whereas undersea cables carry a large fraction of the telephone traffic between pairs of advanced countries, Intelsat is the one system that can connect a person by telephone to someone in essentially every country in the world.

The international satellites are used to feed not only the international television network but domestic networks as well. Foreign news comes to us through these satellites; U.S. programs regularly go to dozens of countries by the same methods.

The Intelsat system has served the world's communications needs very well and is rightly called the pioneer in satellite communications. But it is only the first of a number of international consortia set up to provide satellite services. Next we will discuss its sister organization, Inmarsat.

■ INMARSAT, THE MARITIME COUNTERPART OF INTELSAT

The second best-known international communications system offers services to maritime users such as ships at sea. Inmarsat's creation followed Intelsat's by about 15 years and was brought about by many of the countries that formed Intelsat, with the British playing an organizing role. Inmarsat went operational in 1982, as the result of two separate initiatives, and Comsat can take credit for a strong hand in both. Another important player in the act was IMO, formerly IMCO, the International Maritime Consultative Organization.

In May 1972 Comsat Labs put an eight-foot satellite dish on the ocean liner *Queen Elizabeth II*, to show some of the varied uses of small receivers tied to satellites. Both crew and passengers used telephone and facsimile services supplied by this small terminal. Its success sparked a wide debate about how best to provide satellite services to marine users. Dr. Burt Edelson, director of Comsat Labs at the time and a former naval officer, invited the Chief of Naval Research to see the demonstration on the *QE II*. The admiral was properly impressed and asked if such service could be provided on Navy ships.

To get a maritime system under way, Comsat had conceived a dual-purpose satellite. Part of its capacity would serve the marine industry, using frequencies controlled by the FCC and ITU for maritime use. The rest of its capacity would serve the needs of the U.S. Navy, using frequencies reserved by governments for their own purposes. Comsat called its satellite system Marisat. Comsat knew that inordinate delays in delivery of a satellite system called Fleetsat had created a communications gap for the Navy. Like many military systems, Fleetsat had become overweight and overly complicated. Because it carried several different communications systems, Fleetsat had a plethora of serious mutual interference problems. As these problems were being solved, delays of months dragged into years.

Comsat proposed to the Navy that its immediate needs could be met by leasing capacity on a satellite of simple design, and offered to put up such a system at no cost to the Navy—if the Navy would sign a multiyear contract for service. The Navy was happy to see a ready solution to its problems with Fleetsat and awarded Comsat a two-year contract to provide service in two oceans, the Navy considering its needs in the Indian Ocean too small to justify service there. After two years the Navy expected Fleetsat to come into service. Comsat called their system Marisat, but the Navy called it Gapsat, since it was supposed to bridge the gap until Fleetsat was ready.

Working with Hughes Aircraft, Comsat designed and bought three satellites from Hughes to do the job—two to cover the Atlantic and Pacific oceans and a spare in case either of the other two failed to reach orbit. (Launch failure is the most common problem, although a failure in orbit is not unheard of.) Very quickly the Navy saw the advantage of worldwide coverage and contracted for service in all three oceans. Comsat launched the third (spare) satellite to serve the Indian Ocean and contracted with its Italian partner in Intelsat, Telespazio, to provide facilities for control of the satellite.

The Marisat satellites are all still working reliably since their launch in 1976, even though they were designed for only 5 years of life. These satellites were well constructed, and Comsat has been able to stretch their lifetimes by adroit schemes to conserve battery life and expendable hydrazine maneuvering fuel. Marisat was still collecting Navy revenues as of mid-1990. Comsat signed another multiyear contract with Inmarsat in 1989 to continue to provide Marisat capacity as backup for Inmarsat's commercial users.

Through a series of extensions the Navy's contract with Comsat has spanned a period of 15 years. The Marisat program was a great moneymaker for Comsat, even though the Navy drove some hard bargains, knowing that Comsat had recovered its total investment in the early years.

With three Marisat satellites in orbit to serve the Navy, Comsat was in a position to provide essentially worldwide commercial service to the merchant marine and other private users. To do this, Comsat contracted with KDD in Japan to operate a so-called coast earth station to give access to the Indian Ocean satellite, thus completing the worldwide system. To ensure that potential customers would not be kept waiting, Comsat stocked an initial supply of ship terminals bought from Scientific Atlanta and made them available for lease or purchase. In a few short years, several suppliers of ship terminals entered the market and Marisat was handling traffic from 1,000 customers. Thus the maritime communications system had become a success in its own right.

While Comsat was building up the commercial side of the Marisat business on its own, a number of countries had made known their interest in establishing an international consortium, which became known as Inmarsat. IMCO became the agency through which interested parties (including the United States) exchange information and plan future actions. Meanwhile, Congress amended the Satellite Act in 1978 to designate Comsat as the U.S. member of Inmarsat. Since it took several years for Inmarsat to evolve from a paper organization to a going concern, Comsat continued to manage the Marisat system. As the U.S. member, Comsat lent a hand in various ways to get Inmarsat off and running.

Between 1976, when the Marisat system began working, and 1978, when the Satellite Act was amended to make Comsat the U.S. member of Inmarsat, some 400 ship terminals were installed on customers' ships. These included freighters, cruise ships, yachts, barges, oil rigs, and VIP airplanes. There were also a few fixed sites in places like

Antarctica that wanted service. By the time Inmarsat went into operation in 1982, there were 1,000 customers ready to use its service. That number was 11,000 in mid-1990.

When Inmarsat began operations, it depended on the three coast earth stations established by Comsat to gain access to the satellites. Now there are 21 such stations in operation, reducing the need for signals to travel long distances via other means before reaching the customers in various countries. Almost a dozen new stations are under construction.

Like Intelsat, Inmarsat's success has exceeded its founders' hopes. With 58 member countries, it is a smaller organization than Intelsat; although its revenue is much lower than that of Intelsat, it has an impressive rate of growth, taking in about $74 million in 1987, about $100 million in 1988, and $130 million in 1989.

Inmarsat's governance is similar to that of Intelsat. Members elect a council which in turn selects a director general. The present director general is Olof Lundberg, formerly of the Swedish PTT.

Anyone wishing to go into the space business could well envy the Inmarsat story. By the time it was ready to offer service, many of Comsat's customers were standing in line with their credit cards in hand. Customers who had originally been served by Marisat expected to continue to receive service, and Inmarsat found a ready way to serve them. It leased capacity on three different satellite systems—Comsat's Marisats, the European Space Agency's Marecs, and several Intelsat satellites that carried a "maritime package." Thus Inmarsat got into business without having to buy a single satellite.

Once it was successful, and as its needs for capacity grew, Inmarsat decided to buy its own satellites. In the mid-80s it conducted a competition for construction of the satellites, which British Aerospace won. Hughes Aircraft received a subcontract for supplying the heart of the communications system, the transponders. Inmarsat 2 is to have three times the capacity of its predecessor, though its development has been slowed by the usual problem of mutual interference between its different systems. Inmarsat conducted a competition to select a supplier for third-generation satellites, even before the second has flown. General Electric won the competition. Inmarsat 3 will raise the system capacity by a factor of 10 over that of the second-generation system.

In addition to the various industrial users, cruise ships are now a big source of revenue. More and more passengers have come to realize that while they are captives on a ship going nowhere, they can remain in touch with the world by using Inmarsat services.

Inmarsat is now positioning itself to expand into other areas of mobile communications. Recently the Inmarsat Convention was amended to permit offering aeronautical services, with service beginning in late 1990. Inmarsat also expects to offer mobile land services in certain areas of the world. No doubt these new services explain Inmarsat's desire to increase the capacity of their satellites so significantly.

While the Soviet Union chose not to join Intelsat, it did not want to pass up the chance to join Inmarsat, and signed up early in the game. The Soviets developed their own maritime communications system for domestic use and made it compatible with Inmarsat. At an early meeting of the Inmarsat Council, the Soviets said that their ship terminals could shift from domestic service to Inmarsat service by merely throwing a switch. Several other Soviet-oriented countries also belong to Inmarsat.

■ LESSONS LEARNED FROM INTELSAT AND INMARSAT

Because international space communications is the most outstanding example of a commercial success in space, we might legitimately ask what we have learned from this business that might apply to other kinds of space activity. Unfortunately, the chances are that whatever we learned in one case about space commerce will apply only partially to the next situation we face. Also, the environment has certainly changed, the people are different, and there are many reasons why we should not claim to have learned so much about space commerce after all.

Perhaps the first question we ought to try to answer is: Why was the situation ripe for commercialization of satellite communications at that time? Was it a good time to try to start a business in space? Let me mention some things which seem to have been particularly helpful to our success in initiating a commercial operation in space communications.

(1) There was already a thriving multibillion-dollar domestic communications industry in the United States and other advanced countries. Whatever happened in satellite communications would be superimposed on an industry that was already well established and profitable.

(In analyzing remote sensing and other possibilities in space, we need to think about whether they were firmly established businesses before taking off into space.)

(2) Space communications began as an international activity. After all, we had Ma Bell taking pretty good care of our local and continental traffic. So the gap we recognized was the lack of good, reliable, afford-able links to foreign destinations; as a matter of fact, people had to place calls in advance and sometimes wait for days for service. Other space activities may not have such enormous international potential.

(3) Both the government and the civil sector had reasons for welcom-ing advances in communications. This is another way of saying there was a pent-up demand for better communications. This is not to say that there were millions of citizens waiting to go to the phone and call overseas or that millions of businesses depended on good international tielines. But there were many people who knew they could use better communications and who constituted the core of the demand; later, many other people recognized the advantages of better communication overseas. This leads to a more general principle: In starting up a new business, it is obviously helpful to find a pent-up demand for what one has to offer. If that cannot be found, one must create a demand; if that is not possible, then one has a fundamental problem. Both types of demand (pent-up demand and stimulated demand) have propelled the satellite telecommunications industry.

(4) The government's policy was clearly stated in a public document. There was a strong desire on the part of government officials to estab-lish a sound basis for satellite communications and to create an envi-ronment in which people could capitalize on the large investments the government had made in satellite technology and would have the confidence to invest their own money in it. The Satellite Act created the necessary framework for all this to happen.

(5) The government agreed not to compete with the private satellite industry and has lived up to the agreement.

(6) The government stated that it was a potential customer. Although never the dominant customer for commercial satellite services, its need for communications support during the Apollo launches helped Comsat and Intelsat focus on achieving worldwide coverage as soon as possi-ble. Later the government developed a policy of putting 1/3 of its overseas traffic on submarine cable, 1/3 on commercial satellites, and 1/3 on Defense Department satellites. Traffic generated by the U.S.

government constituted about 5 percent of the total traffic on the international commercial satellite system.

(7) The technology was ready. Although we may think of the early satellites as rather primitive, they were capable of doing the job. The first satellite over the Atlantic, Early Bird, had roughly the same capacity as all three transatlantic cables combined (300 voice circuits).

(8) There was a favorable cost/benefit ratio. Satellites brought an immediate reduction in the cost of transatlantic telephone calls.

(9) NASA's role was clearly spelled out in the Satellite Act. In addition to its contribution to research and development of satellite communications, it was expected to launch the satellites on a cost-reimbursable basis.

(10) The role of various government agencies was outlined. Since the new satellite system was being given a de facto global monopoly and Comsat a legal monopoly in the United States, Comsat had to be regulated to protect the public interest. What agency(ies) would regulate Comsat and what objectives they would work toward were defined in the Satellite Act.

(11) Steps were taken to deal with foreign competition by giving all member nations of Intelsat a role in its ownership and governance, which in turn gave them incentives to protect their interests in the fledgling organization. All of them shared in the liberal patent and data rights policies of Intelsat; some of them would be able to participate in satellite construction. It also kept the doors open to other countries with space interests, encouraging them to join. On the other hand, it also circumscribed what members could do by way of competing.

(12) All interested parties were led to believe that the operating environment would be unchanged for the foreseeable future, providing the stability necessary to encourage long-range planning and investment.

These points, taken together, created the proper atmosphere for getting the first commercial space venture off the ground. Since both Intelsat and Inmarsat are nonprofit organizations, one can raise the issue whether such activities constitute space commerce. This in turn requires that we further refine our definition of space commerce. Intelsat charges its 119 member countries enough to cover its operating expenses and to recoup the cost of new facilities, but it does not make a profit. Its members, on the other hand, have the right to make money on the services they provide to their customers; Intelsat itself is not normally allowed to deal directly with the customers of its members. Comsat, the United States' representative in Intelsat, is a private company, but one that was created at government request and is regulated like an electric utility. Still it is expected to compete in the marketplace with a number of other private companies that have recently sprung up.

Intelsat and Inmarsat are the two best-known satellite systems. But Comsat has played an important role in the acquisition of regional satellite systems by a host of countries with varying levels of sophistication about satellite communications. All told, Comsat has provided technical and management assistance to about 50 countries in designing overall systems and in acquiring satellites, ground stations, or both. Several satellite systems other than Intelsat and Inmarsat now provide services to more than one country. Of these, probably the two best known are Eutelsat and Arabsat.

▪ EUTELSAT

In 1967 France and Germany agreed to build jointly a satellite called Symphonie, the first European experimental communications satellite. Working with a few other countries, the French took the initiative to move aggressively into the space age. They formed the European Space Research Organization (ESRO) and the European Launch Development Organization (ELDO), two agencies which in 1975 were merged to form the European Space Agency (ESA). In 1970 Europe first spoke of having its own satellite communications system when the CEPT (Conference of European Posts and Telecommunications) representatives met in Brussels, and a study group was formed to examine the possibilities. They envisioned that an international body might be established to operate such a system. By analogy to Intelsat, it was named Eutelsat.

In 1973 the nine members of ESRO decided to build the Orbital Test Satellite (OTS), looking forward to developing an experience base which would later enable the construction of the European Communications Satellite (ECS). Soon after the launch of Symphonie in 1974, ESA came into being and the development of ECS satellites was approved. In 1977, 17 members of CEPT approved the formation of an organization called Interim Eutelsat, which two years later decided to buy five ECS satellites at a reduced price from ESA. Shortly thereafter, Eutelsat hired Andrea Caruso, a high official in Intelsat, as its Director General. (Caruso was replaced in 1989 by Jean Grenier who came from the French PTT and had served on Intelsat's and Eutelsat's Board of Governors.)

Being good parliamentarians, the Europeans knew very well that whatever they did by way of creating a commercial (as opposed to an experimental) satellite system would have to be coordinated with Intelsat. While some said the Intelsat Agreement was not clear on this point, what Europe had in mind definitely required coordination.

If there had ever been any doubt, the Europeans' attention had been captured by the debate about launching Symphonie. The United States had refused to launch it until we were assured that Symphonie would not compete with Intelsat for commercial traffic but would remain strictly experimental in nature. This fracas no doubt gave added impetus to the European wish to have ESRO and ELDO succeed, and in fact to have them merged to form a more effective agency, ESA. Their determination culminated in the creation of Ariane.

With the European decision to purchase ECS satellites from ESA, they faced the need to plan how they would go about coordinating the Eutelsat system with Intelsat. Soon they announced their intention to seek coordination under Article XIV(d) of the Intelsat Agreement, the item dealing with the avoidance of significant economic harm and technical interference to Intelsat.

Americans and others pointed out that European traffic, as opposed to the traffic in some smaller regions, was really large enough to be significant, and if drained off by Eutelsat could in fact cause a problem. The European rejoinder was that, as a matter of fact, they had made plans to build a vast ground-based network to carry voice, video, and data among the European countries, but, on further thought, they had decided to cut short those plans and build a satellite system instead. This meant that any traffic carried by Eutelsat would be traffic siphoned off the proposed terrestrial system—not traffic taken from Intelsat!

This carefully planned strategy plus Europe's political leverage allowed them to proceed with what they described as a minimal system consisting of one satellite called Eutelsat I-1. They were able to sell this plan at the Intelsat meeting of April 1979. Even though there were objections, Europe's clout succeeded in achieving coordination—a lesson not lost on the United States and others needing coordination for their own plans, as we will see.

The Europeans were soon back at the table again, this time seeking coordination of Eutelsat I-2, a spare satellite merely intended to ensure continuity of service. It was quickly coordinated in October 1980. Within a year, Eutelsat was back once more, this time planning to put a significant amount of TV traffic plus some specialized business traffic on their second satellite. The European foot in the door had become a pair of boots. Finally, the interim arrangement of 1977 was superseded by the agreement of 1982 and Eutelsat officially came into being.

Following the lead of ESA and the European Broadcast Union (EBU), Eutelsat agreed in 1982 to provide capacity for a TV service called Eurovision. By 1983 Eutelsat was in operation in an all-European mode. Europe had built an ECS satellite, had launched it on Ariane, the rocket built by an all-European consortium, and had put it in service as Eutelsat 1 F-1.

Since then, Eutelsat has prospered. It now operates 4 satellites—all, needless to say, launched by Ariane. Interestingly, most of Eutelsat's capacity is devoted to TV distribution—not to hauling telephone traffic. Like Intelsat, Eutelsat has an executive organ headed by a Director General who supervises an operating staff located in Paris. In fact, the overall organization of Eutelsat is Intelsat in a regional context.

While certain critics have decried the establishment of Eutelsat and its competition with Intelsat—a system that could easily have supplied most of the services Europe wanted—its development was a foregone conclusion. Europe saw Intelsat as a U.S. creation from the beginning and used it as a stepping stone to creating a European carbon copy. European industry cut its teeth making components and subsystems for Intelsat satellites under contract to the prime contractors, Ford and Hughes, both of which have committed to giving a larger and larger share of satellite parts construction to a selected group of European partners. The choice of European partners has come to reflect the voting power of the approving agencies.

Since Canada and Japan have also demanded a piece of the action, Intelsat has become a prime agent for technology transfer from American satellite builders to those who will soon be in an excellent position to compete. This possibility was undoubtedly on the minds of those countries when they decided to join Intelsat many years ago.

The story of how Eutelsat managed to receive coordination from

Intelsat to build a system that is clearly competitive with the original international system is a good lesson in international political science. But diplomatic skill aside, there was clearly a certain imperative in this case to permit a little competition from the burgeoning regional systems like Eutelsat. In fact, there were those who felt that Intelsat—the most successful internationally owned commercial undertaking of all time—was getting old and lazy and needed the stimulus of competition to shake it out of its lethargy. The rules designed to avoid significant harm enabled Intelsat to come into being and to have the assurance of the essentially universal support necessary to its early success. It would be naive to think that the Europeans would forever deny themselves the opportunity to build an autonomous space industry—an industry that in their minds is a sine qua non for their future competitiveness and viability in advanced technology. And the Europeans saw Eutelsat as an essential part of a well-rounded space capability. Once we accept that fact, we can appreciate Eutelsat for what it is: a successful and useful system in which all its partners can take pride—a pride they never seemed able to generate with regard to Intelsat.

Eutelsat's ability to receive coordination from Intelsat made the United States an eager student of the art of successful diplomacy as practiced by Intelsat's European members. For some time the United States had been making plans to expand the area covered by U.S. domestic satellites, and the European initiatives gave the United States the opportunity to come out of the closet and ask to use domestic satellites for service to Canada and Bermuda.

All the major parties seeking coordination of regional systems with Intelsat have used the actions of other groups to justify their own, even though each has felt that it had a certain unique logic on its side as well. Among other arguments, the Europeans said that it was merely a historical happenstance that America was one country while Europe was not: after all, a United States of Europe was just a few years away, and so the situation in both Europe and America should be treated as being essentially similar. This position carried some weight, but in the end, the deal boiled down to Europe's agreeing not to challenge the United States on domestic satellite connections to Canada and several nearby islands, provided we did not challenge them on building a European system that was clearly competitive with Intelsat.

By this means, both parties received the necessary coordination in October 1982.

Although the main justification for Eutelsat was to provide telecommunications among members, it has now also become the principal supplier of television distribution for Europe. Eutelsat now operates 4 satellites of 14 transponders each, a large part of the capacity of which is used for television. More than a dozen different TV services are now being provided. Its satellites also contribute to feeding some 15 million European homes that depend on cable TV services.

The number of countries that are logical candidates for Eutelsat membership has taken an upward turn with recent developments in Eastern Europe. Whereas the most obvious manifestation of the new look in Eastern Europe is a spate of open elections, it could well be that the next status symbol for those countries will be Eutelsat membership. Poland joined in February 1990, bringing total membership to 27. Romania joined Intelsat in May 1990 and is expected to join Eutelsat shortly; also Hungary, Bulgaria, and Czechoslovakia are said to be moving in that direction.

It is a fact that the Eastern European countries are hurting badly for investment capital. If they had the money to invest, they would no doubt spend a great deal on installing modern telecommunications networks, knowing the importance of communications in enabling economic development. Lacking the capital, these countries can make good progress in satisfying internal needs at very low expense by joining Eutelsat—thus benefiting by all the money the partners have already invested.

Eutelsat is now a part of the "establishment," considered by some countries to be too stodgy to keep up with the times and therefore justifying the heretical step of buying capacity elsewhere. Like the members of Intelsat, Eutelsat's individual members are beginning to do things the organization would prefer they did not do. The Astra satellite of Luxembourg has offered services to European countries, in competition with Eutelsat. British Telecom has broken ranks and signed up for 6 transponders on Astra. It has also apparently agreed to be the marketing agent for Astra in finding additional customers. Murdoch's Sky Channel has also taken several Astra transponders, and British Satellite Broadcasting plans to compete with Eutelsat's Europsat; both hope to be the big Direct Broadcast System (DBS) for

Europe. (DBS satellites transmit enough power to the ground that customers may view TV pictures using very small antennas—typical antennas are less than 1 meter diameter.) All these activities are very divisive—some parties justify their behavior in the name of free enterprise, and others say free enterprise must stop short of doing damage to "our" Eutelsat system.

Some people have said that Eutelsat is not really a commercial organization like Intelsat because it got the five ECS satellites from ESA at a reduced price of 60 million EAUs (European Accounting Units), or about $55 million (U.S.). In that sense, Eutelsat is similar to British Airways and Air France, which began operating the Concorde after they were given the airplanes by their governments. It appears that Eutelsat has now reached breakeven, and as traffic grows, as it inevitably must, the enterprise will probably become more profitable.

Eutelsat's satellites are very sophisticated, with multiple beams favoring various parts of Europe. Continuing its drive to remain in the forefront of technology, Eutelsat ordered a second generation of satellites of even greater capability in 1986, the first of which was launched in August 1990. The European satellite industry can be said to have come of age. Not only are they building satellites for their own domestic systems as well as for Eutelsat, but they have bid successfully for Third World satellite production, as we will see in the story of Arabsat.

■ ARABSAT

In the late seventies, Comsat was approached to see if it would like to assist an Arab consortium in acquiring their own regional satellite system, to be called Arabsat. After much negotiation, Comsat received a contract to help plan and build the system. It lasted into the mid-eighties, when two Arabsat satellites were launched: one on Ariane and one on the space shuttle STS (Space Transportation System).

The Arabsat system consists of three satellites (two in orbit and one ground spare), ground control stations, and support facilities. The primary ground control station is located near Dirab, Saudi Arabia, with a backup station near Tunis. The headquarters is located at Riyadh. Individual ground stations in each country are supplied by the countries themselves and are a separate item from the satellites and control stations. This breakdown between the "space segment" and the "ground segment" is similar to that of other regional or international systems where each nation buys its own ground stations.

The Arabsat satellites are unique in frequency choice, having 23 transponders operating at C band and one at S band (2–3 GHz). S band was included because of the original plan to operate a broadcast channel for Pan-Arab television service to all member countries. Although it was expected to be a major offering of the system in the beginning, TV has not become a very important part of Arabsat services, providing only about 10 percent of total revenue. About 70 percent of Arabsat revenue comes from telephone service between Arab countries, with about 20 percent coming from domestic services.

Originally, India's experimental Insat system prototype using ATS-6 (see Chapter 1) was considered the model for an educational series to be offered over Arabsat, but that has not materialized. One should not be too surprised at that development, since most countries are quite jealous of other countries' trying to educate, or possibly indoctrinate, their children. While some people are looking to use the S-band channel for business services of various types, others have not given up on its becoming an educational TV carrier in the future. Faisal Zaidan—chairman of Arabsat's board and vice minister of Saudi Telecom—thinks that the TV channel can bring about a unification of the Arab language. All Arab nations write Arabic the same way, but pronunciation differs markedly from country to country. Saudi Arabia broadcasts its basic TV channel in Arabic to all the member countries, as well as its second channel, which is in English; this may explain Zaidan's hope for unifying the language through TV broadcast.

When Comsat was asked to assist in coordinating Arabsat with Intelsat, we worked with them to determine what level of traffic might be diverted from Intelsat. Early estimates were 7 percent. This was soon judged to be too high an estimate, and it was revised downward to about 2 percent. In view of the slow buildup of traffic, this figure does not appear to be out of line. However, things show signs of changing. In addition to Saudi Arabia, we have seen Oman and Mauritania signing up for capacity to be used for domestic service from Arabsat, with Morocco, Libya, and Algeria joining up some time later. These are countries which have reserved, or were expected to reserve, capacity on Intelsat. The traffic on such channels is being diverted to Arabsat.

When Arabsat was in the planning stage, the Arabs knew that Arabsat would have to be coordinated with Intelsat, and since Comsat was under contract to help implement the system, it fell to Comsat to advise them on how best to carry out the coordination process. Their schedule called for coordination more or less simultaneously with that of Eutelsat, and the two systems were compared with each other in various ways.

The Arabs, taking several lessons from the European textbook, began talking of a vast ground-based network they had planned to build across the deserts of the Middle East and Africa. Like Europe, they said that Arabsat was not actually competing with Intelsat but merely with the landlines they had planned to build to handle traffic growth. They then claimed that, on second thought, it made more sense to build Arabsat than to string lines across the desert, and they probably could save money and lives in the process.

Intelsat members who might have thought these arguments disingenuous had to keep the big picture in mind. On the one hand, a loyal Intelsat member should vote no on letting various people build systems which might do economic harm to Intelsat. On the other hand, we all must live in the real world. To get along, we sometimes must go along. Any Intelsat member who might have wanted to vote no on either of these systems had to think of all the ramifications of such a vote. Many European countries depended on Arab oil, and perhaps a no vote could be misunderstood as being unfriendly to the Arab countries. Similarly, many of those who might have wanted to vote no on Eutelsat were involved with the Europeans on various trade matters, and, again, a no vote might have been misunderstood.

Using arguments similar to the Europeans', the Arabs mustered enough votes to obtain coordination in April 1980, so in the end both Eutelsat and Arabsat were successfully coordinated.

One might ask whether Arabsat is a commercial system or merely a state-owned system. The answer is a matter of definition. British Airways and Air France are considered to be commercial airlines, and Fiat and Citroen are commercial auto makers. Despite the fact that they are state-owned, their agents and employees probably think they are commercial; whether they are is debatable. Arabsat falls into these same murky waters. The owners are the PTTs of the Arab countries, none of which is privately owned. Its lease services are similar to those in Eutelsat.

▪ MOLNIYA

With the running start of Sputnik, one might expect the Soviets to be the first to establish a communications system by satellite, and as a matter of fact they were in business at about the same time we were.

In the 1960s the Soviets were not nearly as impressed with the geo-synchronous orbit as we in the United States were. The principal advantage of that orbit is that ground stations have a very easy tracking job to do; the satellites appear stationary in the sky. One need make only minor tracking corrections during the day as the satellite wanders perhaps a few degrees in location under worst conditions; normally it is held to about 0.1 degree in position. This allows most customers to have their antennas installed and aimed once, operating from then on without any means of tracking at all.

But for a country such as the Soviet Union which has fairly large cities in areas not far from the North Pole, the view of a satellite over the equator is less than ideal; for the northernmost users, the satellite may be as low as one degree above the horizon. This is not always satisfactory for reliable communications.

To deal with their northerly situation, the Soviets used an orbit which in *Voices from the Sky* Arthur Clarke credits to a Britisher named W. F. Hilton. It is the so-called Molniya orbit. This orbit is at a 65-degree inclination to the equator and is highly elliptical. The satellite reaches geosynchronous altitude (22,300 miles high) at the top of its orbit, whereas its perigee is only a few hundred miles high. Its period of rotation is about 12 hours rather than 24, but a satellite in this highly elliptical orbit lingers high in the sky for about 8 hours, moving only very slowly across the sky. If three such satellites are properly phased, they can provide to users at high latitudes most of the advantages of a truly geosynchronous orbit.

In April 1965—the same month that we launched the Early Bird satellite—the Soviets launched their first satellite into the Molniya orbit. Perhaps we should credit them with inventing the orbit, if they did not know of Hilton's work. In any case, they have launched dozens of Molniyas. Unfortunately, continuous service demands there be two or three costly tracking antennas at each ground station. Even so, there are clearly cases where service ought to be provided, even when the economics are not favorable. Those northern sites might get no service were it not for Molniya. (The United States provides a number of services to the islands of the Pacific in spite of the small amount of traffic they generate and in spite of their lack of economic viability.)

About 10 years ago the Soviets decided the geosynchronous orbit was interesting after all, and began to plan for Statsionar satellites, which would go into that orbit. They filed with the International Frequency Registration Board for a number of orbital slots, and since then have launched four or five satellites into geosynchronous orbit.

■ INTERSPUTNIK

In the early sixties when Intelsat was coming into being under the organizing influence of the United States, our State Department and Comsat were sincere in not wanting to exclude any country from taking out membership in Intelsat. Consequently, the Soviets were invited to come to the organizing meetings. They sent representatives but in the end chose not to join. Observers have postulated various reasons why. Certainly they would have benefited from being in a position to oversee the development of U.S. communications technology. The most likely reason seems to be that they would have been embarrassed to let us see how little traffic they generated. We know that, even now, the Soviets' use of communications is minuscule compared to most Western nations. In 1960 it would have been even smaller.

But having chosen not to join, and having thus set a pattern which their allies would have to follow, they decided to offer their allies service over a system they called Intersputnik. This system was not like Intelsat. It did not—and still does not—have any satellites of its own. Rather, it leases time on a few transponders on Soviet satellites normally used for domestic service. Using Molniya in the beginning, it now also uses Statsionar satellites.

About five years ago, I visited the Intersputnik offices in Moscow and was given a briefing on the system. My host was the Director General, Spartak Kurilov, who remained in office until 1989. (He was replaced by Boris Chirkov of the ministry of telecommunications.) From information Kurilov gave me, I concluded that Intersputnik generates only about one percent as much income as Intelsat does (now $600 million per year). In fact, at last count, the system supplied only about 300 circuits to its users, plus some television. This compares to 120,000 circuits for the Intelsat system. Intersputnik has less than 20 members, and a typical member generates very little traffic.

About a year later I was offered a tour of an Intersputnik ground station near Budapest and eagerly accepted. It included a large dish about 60 feet in diameter on top of a building which contained the communications equipment. The basic equipment appeared to be sturdy and well built but about 25 years out of date. I took photographs of everything in sight until the guide told me that they did not allow people to take pictures because they did not want them to fall into the hands of the CIA! Perhaps with today's *glasnost,* this would not have been an issue.

The Budapest site was manned by about 25 people. We were told that the site was the terminal for about 40 circuits. The traffic was only a few percent of what one would have expected from a similar site in the United States or Western Europe. Judging by the figures given to me by Mr. Kurilov and by the station manager in Budapest, it seems obvious that the Soviet system is not carrying enough traffic to be commercially viable and must be supported by the government for purely political reasons.

Intersputnik's members are countries of Eastern Europe as well as Cuba, Nicaragua, Vietnam, and other underdeveloped countries, but there are no members in the sense of ownership, since there are no commonly owned satellites. Intersputnik leases capacity on Soviet-owned satellites.

Like Eutelsat and Arabsat, Intersputnik should probably be classi-fied as a regional system. It is certainly regional in a political sense. Intelsat has no political litmus test and is open to all members of the ITU who are willing to accept its charter. Members include communist countries like China and Vietnam and indescribable economies like Iran as well as the full range of capitalist countries from Sweden to the United States.

Intersputnik does have a litmus test, in that member countries are those in the Soviet economic orbit. A few members of Intelsat such as Algeria have chosen to join Intersputnik in order to have connec-tions to several additional countries. Conversely, some Eastern Euro-pean countries are already joining Eutelsat, and Romania has already joined Intelsat; it probably won't be long before Poland and Hungary will apply for membership in Intelsat. The Soviet Union has indicated that it wants to improve relations with Intelsat, if for no other reason than to simplify the inevitable problems of signal interference between the two systems. The Soviets will no doubt try to delay any more Intersputnik members from defecting to Intelsat until it sets up its own arrangements with the Western-oriented system. Having them as members would drive another nail in the Cold War coffin.

▪ U.S. COMPETITION WITH INTELSAT

Early in this chapter, I referred to the lengths to which the founders of Intelsat had gone to control the level of competition. The final agreements called merely for consultations, to ensure technical compatibility and to avoid "significant economic harm" to Intelsat. Eutelsat, Arabsat, and Indonesia's Palapa group were all successful in passing the Intelsat coordination hurdle, as was the United States in its bid to serve non-U.S. islands such as Bermuda and to provide certain limited services to Canada and Mexico.

Perhaps more important to the health of Intelsat was a U.S. decision to permit competition by companies wanting to offer commercial satellite service across the Atlantic. In the early eighties, this subject engaged the attention of the FCC, State Department, and many others. Since the United States had been the main proponent of Intelsat in the beginning and had come to its defense on many occasions, this was quite a new departure in terms of policy. Many former government officials—not to mention Comsat executives—were horrified at the thought that we could somehow compromise our earlier loyalties.

Finally, the United States took the position that competition by private companies would be permitted provided they were offering new kinds of service—something other than public-switched telephone service. This meant they could not interconnect with the public-switched network. They could offer TV connections, data services, and the like; they could even offer telephone service if it were private line only, such as service from a General Motors plant in the United States to another in Europe.

As a first step, it was necessary for a potential offerer to find two countries who would go to Intelsat and ask for service. Several companies have availed themselves of this opportunity. Following its success in getting the Peruvian and American representatives to ask for service, PanAmSat has been successfully coordinated to provide certain special services to Europe and Latin America. Its president, Rene Anselmo, is a vocal proponent of privately owned satellite systems and now provides private services to a number of countries. To get

Figure 7. Motorola, taking advantage of the new U.S. policy toward competition with Intelsat, has announced that it will develop a worldwide system of telecommunications which it calls Iridium. It is based on a constellation of 77 satellites in low earth orbit that use cellular technology to communicate with users on earth. The signal is transmitted from the caller's lightweight, portable phones directly to the nearest overhead satellite, which in turn sends the signal to an earth "gateway," which verifies the caller as an authorized user. The call is then routed through the constellation of satellites to its destination anywhere on earth. The first launch is scheduled for 1992. (Photograph courtesy of Motorola Satellite Communications.)

started, PanAmSat offered some very attractive rates and, ironically, later objected to Intelsat's cutting its own rates to be more competitive with PanAmSat.

In 1989 another U.S.-based company, Orion, placed an order with a consortium headed by British Aerospace for a turnkey system including two Ku-band satellites delivered in orbit on the Atlas rocket. They propose to begin operations in 1993 to provide various nonpublic network services—especially VSAT networks.

So far, such systems have caused no damage to Intelsat and may have strengthened it. The fear of competition can certainly cause an organization to shape up and deliver better service to its customers; Intelsat's response to PanAmSat and Orion seems to bear out this principle.

3

DOMESTIC COMMUNICATIONS SATELLITES

The need for a better international communications system was the driving force behind the establishment of Comsat and the creation of Intelsat. Once Intelsat was operating smoothly, Comsat turned its attention to growth opportunities in domestic satellite communications. It sought approval from the Federal Communications Commission for a Comsat-owned system dedicated to domestic use.

The United States, like many other advanced countries involved in the creation of Intelsat, had a rather good land-based domestic communications system already in place. And in those instances where satellites were needed for domestic communication—for example, from the continental United States to Alaska, Hawaii, and Puerto Rico—service could be obtained from Intelsat. Thus there appeared to be no particular urgency about approving domestic satellites (or domsats, as they came to be called). Consequently, the Nixon Administration and specifically the FCC took its time and conducted several studies of the pros and cons of different methods of achieving domestic satellite services.

What it eventually concluded was that, because the existing system appeared to be meeting the needs of users, neither Comsat nor anyone else should be granted a monopoly on domestic service. Adopting an "open skies" policy, the FCC told interested companies that they would be considered for licensing if they could show that their plans fulfilled some social purpose to justify use of the spectrum and if they could demonstrate the necessary technical competence, legal status, and financial resources to put together a system. This policy has been applied to other domsat services, such as direct broadcast. (The situation with respect to mobile services will receive special treatment later in this chapter.)

In the first decade following the establishment of the ground rules, a few companies did accept the challenge, and as a result 12 domsats were put into orbit in the 1970s. These early domestic systems went through some startup problems because of a very limited demand for capacity. After 5 or 6 years, however, a thriving domestic satellite business emerged, and the owners of the original 12 domsats put up an additional 12 in the early eighties. Based on these perceived successes, a second round of interest developed about a decade after the original companies had taken the plunge. It appeared that the demand for domsats was growing by leaps and bounds, with no end in sight. If there was money to be made, there were people who wanted to put money at risk to make more. A surge of interest swept a number of companies into the field. When the FCC notified the industry in the early eighties that it was accepting applications, a number of companies asked for approval to launch about 30 satellites of various types.

The FCC was overwhelmed at the response. They said that they might have to resort to a lottery to resolve the question of just who would be authorized to launch domsats. But as systems already approved became operational, capacity developed faster than users could absorb it. The satellite market became overbuilt—somewhat like the commercial real estate business (both being stimulated in part by the favorable tax policies of the Reagan years). As people became more aware of the highly competitive nature of the domsat business, many who had asked for authorization to put up systems returned to notify the FCC that they had changed their minds. These cancelled plans simplified the FCC's problem in deciding among applications, and so no one who wanted to participate was denied the opportunity. Of the 30 or so domsats that were expected to be launched in the eighties, only about half were actually launched—partly because some of the interested parties got cold feet, and partly because of the unavailability of launch rockets after the loss of the space shuttle *Challenger*. Counting the ones that were eventually launched and the 15 or so that were already operating in orbit, we now have about 25 domsats in service (in various stages of health), owned or operated by seven U.S. companies. In addition, there are four Canadian and two Mexican domsats that also supply service to American users.

The FCC monitors transponders on domsats to see if they are in use and publishes the results. There are 300–350 transponders out there looking for domestic traffic of one kind or another. In 1988 the FCC noted that only about half the existing transponders were carrying enough traffic to be considered economically viable. However, since that time, growth of demand has improved the prospects for transponders and there are now satellite providers clamoring for a means to launch additional capacity. The commercial launch industry (see Chapter 4) is in the process of attempting to satisfy those domsat owners who want to become customers for their services.

The experimental work carried out by NASA and the military, and the successes of Intelsat, led the FCC to the conclusion that domsats, to be practical, should operate at either C band or Ku band and, in fact, should look a lot like the satellites that had been developed for Intelsat. The main difference between a domsat and an international satellite is that the typical domsat confines its energy to one country, whereas the typical international satellite spreads its energy over about one third of the earth's surface. This means that international service requires rather large antennas on the ground to pick up enough power from the satellites for satisfactory service, whereas domsats can get by with much smaller "dish" antennas.

But what new services were desirable or necessary? The FCC believed that it had to make those decisions, keeping in mind its guardianship of the airwaves and the presumed shortage of radio spectrum. While the FCC was taking its time in coming to decisions about domestic satellite regulations, our neighbor to the north, Canada, became the first nation to put a domestic synchronous satellite communications system into operation.

■ CANADA'S ANIK SATELLITES

Canadian planning had begun early. In 1969—the same year that Intelsat established a worldwide system and was thus able to telecast globally the first astronauts landing on the moon—the Canadian Parliament created Telesat, a government corporation created to bring a domestic satellite communications system into being. Because of the effectiveness of Canadian planning, only a little more than three years elapsed between the formation of Telesat Canada and the first launch of a domestic satellite.

Communications in the southern part of Canada were as good as those in the United States, but in the vast reaches to the north her communication system was very poor. Domsats were the obvious answer to this problem. Sharing those northern latitudes with the Russians, perhaps Canada could have been a candidate for the Molniya orbit used by the Soviets. But while Molniya does allow satellites to linger over the northern latitudes, they do nevertheless move out of position after a few hours, and so more tracking and switching is required than for satellites in the geosynchronous orbit. Therefore the Molniya orbit does not lend itself to simple ground stations—an important point for use in small, isolated Canadian villages. Consequently, Canada became the first country to operate a domestic satellite system using geosynchronous satellites. NASA launched the satellites, rubbing a bit of salt in American wounds, since we had no domsats of our own at the time. (The FCC can take some of the credit for that lag since it had taken six years to complete the rulemaking process.)

Canada's first domestic satellites were launched mainly to unite its citizens in northern areas with those in the rest of the nation. The Canadians called their domsats Anik (Eskimo for brother). The first message transmitted was a call from a small Eskimo village, Resolute, to Ottawa, establishing a permanent link from the remotest village to the nation's capital. Anik 1 went up in late 1972, Anik 2 in early 1973, and Anik 3 in mid-1975.

Canada has continued to lead the field in domestic satellite systems. The first series of Anik satellites was developed by Hughes Aircraft, an American company already well respected for its work in developing satellites for international use. These Aniks operated in the 4–6 GHz frequency range (C band) and were inherently very simple in design. Normal practice is to use 6 GHz as the up frequency—that is, ground to satellite—and to use 4 GHz as the down frequency. An experimental satellite called Tacsat that Hughes had built for the Air Force had used a so-called despun antenna, so that as the satellite rotated, the antenna mounting rotated in the opposite direction, causing the antenna to point constantly downward. Hughes incorporated this feature into its first domsat design and sold it to the Canadians.

In 1978 Telesat, the Canadian satellite communications company, ordered a series of three satellites designed to operate in the 12–14 GHz range of frequencies (Ku band), thus providing higher capacity than the C-band satellites. The first was put into orbit from the space

Figure 8. The satellites of the Hughes Aircraft Company have met the communication needs of the world since Early Bird in 1965 (far left). This geosynchronous satellite was less than 1.3 meters high and carried 240 simultaneous telephone conversations. The HS 333 (3.4 meters high; second from left) carried 6,000 calls, whereas Intelsat 4 (5.3 meters; third from left) carried 9,000 calls. The HS 376 (6.7 meters) carries 25,000 calls, and Intelsat 6 (11.7 meters) carries 120,000 calls. Hughes' newest satellite, the HS 601 (25 meters long; far right) will meet the demand for high-power satellite services, such as direct television broadcasting and mobile communications for cars, trucks, and trains. (Photograph courtesy of Hughes Aircraft Company.)

shuttle in 1982, becoming the first commercial satellite to be launched from the shuttle. The other two went up on the shuttle in 1983 and 1985 and have become the backbone of Canadian coverage for various domestic services to remote areas as well as areas requiring high-capacity service. These include television distribution to local broadcasting stations, cable TV, telephone service, and radio. Satellites in this series feature spot beams tailored to provide high signal strength in those areas with the largest communications needs (the southern region), with some beams suitable for direct broadcasting.

▪ WESTAR AND SATCOM

The first American company to go into the domestic satellite communications business was Western Union. They did so by buying three of the same family of Hughes satellites that had worked so well for the Canadians—two to put in orbit and one spare to be held on the ground in case of launch failure.

The first Westar was launched in April 1974, the second six months later. Although these satellites offered video, audio, and data services, Western Union planned to use them mainly for message service, counting on being their own customer for much of the satellites' capacity. Unfortunately, not much demand for Western Union's message service developed, and the company fell short of the revenues they needed to break even by about $100 million. But beginning about 1978, Westar began to pick up more and more cable television traffic. To accommodate this new demand, over the years Western Union has bought a total of five Hughes C-band satellites. The original three Westars have passed out of service; of the two still operating, Westar 5 carries a heavy load of cable TV service and also carries much audio subcarrier traffic. The Public Broadcasting System leased four transponders on Westar 4 for radio and television service—thus becoming the first network to distribute programs by satellite.

RCA followed Western Union into orbit in 1975 with its own domestic satellite, called Satcom. But because RCA had leased some capacity on Canada's Anik 2 and made it available to commercial customers in the United States before any other service supplier was on the air, RCA claimed to be the first U.S. carrier to offer domestic satellite service.

RCA shared with Western Union the lean times of the early days of

Figure 9. The current and largest member of the Satcom series, the Astro 7000 satellite. The area of its solar array is 535 square feet and its projected propellant life is 12 years. (Photograph courtesy of GE Astro Space.)

domsats. Like Westars, Satcoms were very lightly loaded and caused RCA to lose money until the fantastic growth of cable television bailed out the domsat business. Cable TV quickly became—and re-mains—the largest user of domestic satellites.

In contrast with the Hughes spinning satellites, Satcoms were de-signed not to spin. RCA claimed that this feature improved design flexibility. Their 3-axis stabilized satellites, rather than having an inher-ently high moment of inertia about a spinning axis, develop stability through an internal momentum wheel that turns at high speed, the speed of the drive motor being controlled to make fine corrections in satellite orientation. To keep the motor speed within the proper range, the momentum is "dumped" when necessary by firing small rocket thrusters on the periphery of the satellite.

Both spinners and body-stabilized designs have advantages. Because of their high moment of inertia—they are really spinning gyroscopes—the spinners are considered more stable and can be brought back into service more quickly when they point away from the earth for any reason; they also have fewer thermal problems because their spin spreads the sun's heat energy more uniformly over their bodies. The Satcom design's strong point is that it permits pointing the solar panels orthogonal to the sun to gather maximum solar flux. It is not surprising that Hughes has normally stressed the advantages of the spinner configuration, while RCA normally has stressed the advantages of their satellite's being stationary.

RCA's Astro Division has continually made improvements in the reliability and performance of its Satcoms—Satcom being the first satellite with solid state transmitters. These high-power transistors enabled RCA to predict a useful life for their satellites of 10 years. Then in 1986 RCA was bought by General Electric, RCA Americom became GE Americom, and GE's satellite division merged with RCA Astro to form GE Astro Space Division. Before buying RCA, GE had not been successful in selling commercial communications satellites, but they did have a good government business with the Pentagon and NASA. Orders from the Defense Department, NASA, and NOAA have helped to form a much larger business base than RCA would have had through commercial business alone. GE Americom now has seven domsats in operation—five C-band and two Ku-band—and continues to improve its satellite operations.

Today, Americom's Satcom K-2 is the most powerful satellite in the sky, transmitting 45 watts of power per transponder. (New Hughes designs will be more powerful still.) Eight transponders are used to carry NBC and NBC Skycom television programming, a service which of course demands the very highest reliability. In addition to being a premier carrier of both network and cable TV, and handling not only NBC but ABC and CBS traffic also, Americom carries more radio programming than any other system. With seven satellites in orbit, GE Americom ranks among the top two companies in the volume of transponder capacity leased to customers, with Hughes as its other main competitor.

▪ COMSTAR

The third entry in the U.S. domsat business was Comsat, with their Comstar satellites—the total capacity of which had been leased to AT&T. The first two were launched in 1976 to supply service to AT&T for PicturePhone—a system which allowed telephone talkers to see each other on miniature TV screens. While Comsat was launching the Comstar satellites, AT&T was building ground stations to work with the satellites—huge antennas of 100-foot aperture and costing $5 million each. AT&T planned only seven large antennas to serve the whole country; it had intended to pipe signals to most of its customers over landlines rather than putting dishes close to the customers, as is routinely done today. In retrospect, we can see that this was a huge mistake because customers appreciate the savings brought about by not having to pipe their signals over ground links. New technology—especially the opening of the Ku band and the concomitant feasibility of smaller antennas located on customers' property—made AT&T's huge antennas look like dinosaurs. AT&T also planned to use Comstar for some ordinary telephone service—probably to see if it could maintain quality using satellite links. In this regard, it planned to use Comstar for one-way service only, so as not to incur satellite delay on both legs of a round trip call.

PicturePhone did not take the world by storm, probably because the image on such a tiny screen (about 3 by 4 inches) was unsatisfactory and because the price was so high. PicturePhone was soon abandoned. Fortunately for Comsat, AT&T had signed a contract under which it paid full price ($1.1 million per month per satellite) for the satellites whether it used them or not. Thus, Comsat became the first U.S. company to actually make a profit on domestic satellites.

AT&T later tried video teleconferencing using conference rooms at strategic locations, but this too did not prove to be a bestseller, primarily because of high costs and the need for customers to go to a facility remote from their place of business. Since 1988, however, video teleconferencing is showing encouraging signs of life. Improvements in the technology are bringing costs down and making teleconference rooms within corporations more practical. PictureTel, for example, uses a computer to reduce the bandwidth necessary to carry TV pictures to remote users. This cuts the cost of the links used for video teleconferencing by a large factor. The cost of transmission capacity for teleconferencing, which a decade ago was about $1,000 per hour, has been reduced through the use of PictureTel's computer to a cost of $14 per hour—a reduction by a factor of seventy. Admittedly, at the lower bandwidth the picture quality is not as good, but the message certainly gets across. And users of video teleconferencing receive a far more appealing product today than PicturePhone ever provided, primarily because today's video teleconferencing uses normal TV screens rather than the tiny PicturePhone screen. Such changes in quality and in the cost of links needed for video teleconferencing will probably soon make the service very competitive and lead to rapid growth in the industry. (Compression Labs—a competitor of Picture-Tel—offers a similar service.)

Comsat later launched two more Comstars, and AT&T eventually loaded the satellites with various kinds of traffic. Comstar turned out to be a good investment for Comsat, even though many felt Comsat should have held out for FCC approval of a broader range of services. Once Comstar was approved, the FCC had told Comsat that it would not extend Comsat's monopoly by approving a multipurpose domestic system to be leased to AT&T's competitors. Of course it would not have been a monopoly, since the FCC was granting approvals to others to operate domsats.

▪ TELEVISION BY SATELLITE

In 1951 AT&T provided the first cross-country link of television stations by coaxial cable. Since then, it has provided the interstation connections via microwave links. In 1973 Comsat tried to sell NBC on the idea of connecting its network through satellite service, but the attempt ended in failure. The negotiations were a great success for NBC, however, because to keep a good customer, AT&T reduced its charges to the network by 30 percent. Comsat tried to sell satellite services to CBS and ABC as well, but there were no takers at the time. (ABC had filed one of the earliest applications for domsat service back in 1966, but of course the FCC was just beginning its long soul-searching process at the time.) NBC's decision not to opt for satellite links in 1973 left the door open for the Public Broadcasting Service (PBS) to become the first network on satellite feed when it signed up to lease three transponders on Westar five years later. To pay for this new set of links, PBS received a major grant from the Corporation for Public Broadcasting (CPB), which also helped tie all the National Public Radio (NPR) stations together by satellite. PBS has found satellites to be a very good way to handle its connections. Each station can record programs for later broadcast and show programs live at the same time. This favorable experience led other networks to follow suit.

Comsat made a second trip to NBC in 1983, and this time it made a sale, getting the contract to feed NBC network TV programming to the NBC-affiliated stations by satellite. It originally used Satellite Business Systems (SBS; see below) satellites for the purpose; but NBC, being an RCA subsidiary, preferred to use RCA satellites, and Comsat began using Satcoms as soon as any were available.

Beginning about 1978, cable systems seemed to catch on to the desirability of bringing a much larger variety of programming to their audiences using signals from satellites. Pay-TV services such as Home Box Office (HBO) were some of the earliest users of satellites for distributing their programs to cable systems. A few TV broadcast stations such as Ted Turner's station in Atlanta decided to take their signals outside their normal regions of coverage. Once the feasibility of doing so by means of domestic satellites was appreciated, such distribution became very popular, and both RCA's Satcoms and Western Union's Westar satellites soon were oversubscribed for cable service. Cable TV thus became the savior of domestic satellite systems. About 300 domestic satellite transponders are now occupied by programming tailored principally for cable systems.

The availability of such signals also led to the growth of the "backyard antenna." Because the power of domestic satellites is comparable to that of satellites designed for international use, the signal strength from domestic satellites is typically much greater, since their energy is not spread over an entire hemisphere. (International satellites have also begun to offer spot beams with high signal strength for special services, but normally their signal is less intense, demanding use of larger antennas on the ground.) Adequate picture quality from domestic C-band satellites normally requires an 8-foot-diameter dish.

Backyard antennas became available for the mass market about 1980 and have been bought by about 3 million households. They thus constitute a significant market for television in their own right. They have also been seen by pay-TV suppliers as a threat, because they are not connected to a cable system that can serve as a billing agent for pay-

TV suppliers' products. As a result, most pay-TV suppliers have begun to encrypt their signals, and the backyard antenna owners must rent decryption devices to get signals from HBO and others.

▪ DIRECT BROADCAST SATELLITES

In 1981 Comsat started a company called Satellite Television Corporation with the intention of operating the first Direct Broadcast Service (DBS) system in the United States. At that time, satellite receiver technology had not progressed as far as it has today; also video cassette recorders were not nearly as ubiquitous as they are now. Backyard antennas were very rare, and their signal quality appealed mainly to those who had no other options. The term "direct broadcast satellite" means a satellite whose signal is strong enough to be picked up directly by small antennas at homes and offices. Other satellites require very large antennas on the ground—antennas too large for home use. Cable TV companies can use large antennas with no difficulty to receive signals of high quality to be fed to their customers.

Comsat ordered two satellites from RCA that would allow use of very small antennas—approximately 3 feet in diameter—at the homes of subscribers, and began looking for other partners to share the $500 million cost of putting such a system into place. A number of false starts were made, with various partners backing out at the last minute as they faced the prospect of putting up a few hundred million dollars to create the new system—not knowing if enough people would beat a path to their door.

One reason those potential partners backed out was that they saw an ever-changing scene on the home front. More and more people were buying 8-foot backyard antennas; soon they numbered in the millions. And it was not apparent that STC would have any special advantages over those with whom they would compete. The main source of programming was the people who were already supplying programs to regular network TV or cable TV. Only customers who for one reason or another could not pick up the signals already on the air could be expected to sign up for DBS. While there were some such people out there, their numbers seemed to be shrinking as cable TV continued to grow, as backyard antennas continued to sell, and as VCRs continued to proliferate. Many of the otherwise willing partners for Comsat's DBS venture thought that the combination of factors nibbling away at the potential DBS market was just too much of a hurdle to overcome. Each in turn backed away from DBS.

A presumed advantage of DBS was its ability to operate with a 3-foot dish as opposed to an 8-foot dish typical of backyard antennas. For those in close quarters such as apartment houses, this could be a real advantage. But more and more apartment houses were being wired with community systems or were tying into existing cable TV systems. The small dish was not much of an advantage to people in rural areas with space for an 8-foot dish. Moreover, companies like Hughes had begun building domsats with more power, going from 10 to 20 to 40 watts, and this blurred the distinction between "ordinary" satellites and DBS satellites, since more and more people could get good signals with backyard antennas. The result was that Comsat gave up on the DBS business without ever having gotten started. DBS is a second example where Comsat took a massive approach to entry into a market, only to see the market nibbled away by gradual changes in the competitive picture.

Comsat's difficulties with DBS took place at the same time that there was a strong press barrage about the burgeoning DBS systems in Europe. Both France and Germany were said to be putting up DBS systems that would attract millions of customers. The French were working on satellites called Telecom and TDF-1; using exactly the same satellite, jointly built by both countries, the Germans were developing TV SAT. Both were said to be under construction and would soon be offering service.

Some 8 years later, what has happened to DBS? *Space News* of November 13, 1989, reported that the French and German systems are in service and gradually building customers. British Satellite Broadcasting (BSB) launched its first satellite in August 1989, with service beginning in 1990. Sky Television, operated by Rupert Murdoch, uses transponders on Astra satellites owned by Système Européen des Satellites (SES) of Luxembourg. Sky is said to be growing rapidly, with one million customers. SES is said to be breaking even after only one year of service. Spain and Sweden are also reported to be nearing readiness for DBS system operation and Eutelsat will soon have its Europsat.

Although Comsat decided to drop out of the race, it is interesting to see that work done by Comsat on the so-called flat plate antennas is finally paying off for someone. The Europeans are ordering these antennas, which are flat plates measuring one or two feet on a side that can fit neatly in places where a large dish either would not fit or would appear unsightly.

It seems the Europeans will be able to make a success of DBS where Comsat could not, for a number of reasons. An obvious factor is that parts of the European "boom" are financed by government, which means the system does not have to make a profit to survive. Also, the United States offers a large variety of other options to customers. Cable penetration is very high here, reaching more than 50 million homes, or over half the market. By contrast, even though it is growing nicely, cable is just beginning to make itself felt in Europe (with something like 15 million homes wired). And the interest in DBS in Europe is in many ways just an updated version of the earlier U.S. interest in the backyard antenna.

In the early months of 1990 DBS began showing more and more signs of becoming a significant reality in Europe. BSB was reported to be discussing combining forces with Atlantic Satellites, Ltd., of Dublin. According to *Space News* of March 19, 1990, these two companies are concerned about the headstart of Rupert Murdoch's Sky Television. Sky has signed up for capacity to offer 16 channels of service over Luxembourg's Astra satellite. The combination of BSB and Atlantic would provide 10 TV channels. BSB's success has certainly not been assured, but joining Atlantic could make its prospects look a lot brighter. Atlantic is 80 percent owned by Hughes Communications, which has become a large factor in satellite TV in the last decade, as described below. Hughes' growth has paralleled Comsat's effective disappearance from the scene.

Space News carried an update on August 27, 1990, when it reported the successful launch and checkout of TDF-2, the second French DBS satellite (the first was launched in October 1988). BSB also had a successful launch of its second satellite in August. SES expects a second Astra satellite, with 16 more transponders, to go into orbit in February 1991 alongside its first; viewers will receive 32 channels of video on an antenna about 2 feet in diameter. Already more than 1.3 million Astra-compatible dishes have been sold. BSB expected to have about 400,000 dishes in place by yearend 1990. With the first Eutelsat 2 launched in August 1990, there is much DBS activity in Europe. Eutelsat said that it had sold 52 channels—mostly for TV—on its 5 Eutelsat 2 satellites which will be in orbit by yearend 1992 with a total of 80 channels. About 80 percent of their capacity is slated for TV broadcast.

In the United States several consortia have recently announced their entry into direct broadcasting. One such group, Sky Cable, includes Hughes Communications as the satellite supplier, with NBC and Murdoch's news and TV empire as members and also as sources of programs. Together with Cablevision Systems, they plan to go into business by 1993. Their announced markets are rural areas of the United States which probably will not be cabled any time soon because of low housing density. Because of Sky Cable's ties to the cable industry, many cable systems are probably planning to sign up for part of Sky Cable's programs, if this new DBS system comes into being.

The flurry of interest in DBS in the United States about 10 years ago foundered because not enough people were convinced that a DBS system based on that technology could make money. One reason it had little appeal was that it proposed to offer 6 channels, whereas many households had cable with a dozen or more channels. In sharp contrast, Sky Cable has announced it will have 108 channels of capacity and will operate at a power level permitting customers to use a two-foot receiving antenna. Sky Cable might offer 108 channels of TV by having 3 satellites of 12 transponders each, with each transponder carrying 3 TV channels, or it might have 9 transponders per satellite, with 4 programs carried by each transponder. (The latter is what they announced, but that technology is just now in process of being developed; I would place my bets on the former.)

A second group called K prime includes GE and a number of program providers. Since its satellites are of lower power than those to be built by Hughes, it will no doubt require a larger antenna at the customer's site. Nevertheless, both of these new systems would have two advantages over the planned systems of 10 years ago: enough power to increase markedly the number of channels offered and to permit smaller receiving antennas, and enough sources of programming in the consortium that there should be no lack of something to broadcast.

▪ HUGHES GALAXY

It is not easy to trace the economics of satellite use—or of other communications media, for that matter. The basic cost of the service to the user is often so small compared with what it enables the customer to do that a satellite manufacturer may decide he is in the wrong end of the business. Based on that reasoning, Hughes Aircraft decided to try providing satellite services directly to customers rather than just making satellites for sale to someone who would retail service to the final user. It set up a subsidiary called Hughes Communications, whose principal business is leasing capacity to satellite customers. (It will also sell transponders to customers who would rather own than lease.)

Hughes operates three satellites called Galaxy 1, 2, and 3. With 72 transponders available, Hughes Communications has quickly garnered a major share of cable TV services and appears to have its satellites almost fully utilized—quite a feat in a field where there is a large amount of unused capacity. Recently, Hughes Communications bought three Westar satellites and has marketing rights on three SBS satellites. All these satellites were built by the parent company, Hughes Aircraft, and were no doubt sold to their original owners for a lot more money than Hughes Communications paid to buy them back. Having these satellites has made Hughes a formidable rival of GE Americom—Hughes may even have moved to the number one position in terms of traffic.

In addition to these purely commercial satellites, Hughes operates three of what might be called pseudocommercial satellites. In the early 1970s the Navy ordered a system called Fleetsat that ran into problems and delays. The Navy leased capacity on Comsat's Marisat in the meantime to meet their short-term communications needs. Eventually, Fleetsat was built and operated very well. Later, as the Navy began to plan for Fleetsat's replacement, they remembered fondly their ability to get service on Marisat (or Gapsat, as they called it) without having to buy the satellites. They tried the same technique to acquire capacity to replace Fleetsat on a system they called Leasat. Hughes won the contract and designed Leasat to be launched on the shuttle.

The decision to use the shuttle was ironic, because one of the special advantages of leasing rather than buying is that the customer lets the supplier worry about possible launch failures, satellite problems, and any stoppage of service. But here the Navy had bought service on a system that another part of the government could not launch. The Navy suffered through a four-year delay in the shuttle's first launch, and a subsequent three-year hiatus after the *Challenger* accident. Eventually Leasat was launched, giving Hughes a pseudocommercial satellite program in addition to their Galaxy fleet.

Another recent Hughes move has made it a formidable force in the overall communications market. It bought from M/A-Com what had been Digital Communications Corporation (DCC), which was started by Comsat alumni many years ago. This is now renamed Hughes Network Systems and makes a line of small earth stations called VSATs, which in conjunction with satellite capacity may be the fastest growing market in satellite services. VSAT refers to Very Small Aperture Terminals. The size is important in that small size permits installation almost anywhere in a city, even a rooftop downtown. VSAT also implies that frequencies are made available which permit operating in the congested downtown areas in cities where until very recently satellite communications were impossible because you could not get authority to use the proper channel. The available C-band channels were already in use for ground communications, and to use them for satellite service would have created interference with an already existing service. The opening of the Ku band of frequencies in the early eighties changed all that, and now there is a growing business based on using Ku band at downtown locations.

Based on its new well-rounded capability, Hughes now offers a wide range of services: building a satellite, arranging to put it in orbit, operating it, leasing capacity on it, selling a transponder on it, providing a control station to operate it, and supplying a ground network of VSATs at all outlying stores, plants, and warehouses.

▪ SATELLITE BUSINESS SYSTEMS

Other domestic satellite systems have been developed to offer purely business-oriented services. An example of such a system was Satellite Business Systems. Comsat wanted to create a company to offer such services in the early seventies but was told by the FCC that if it wanted to proceed with Comstar for AT&T, it would have to find other partners to share ownership of SBS. After a false start with one group, which included Lockheed and the fledgling communications corporation MCI, Comsat set up a joint venture in 1975 with IBM and Aetna Life Insurance as partners. It planned to offer a private all-digital communications service to corporations, many of which saw their communications needs expanding rapidly at that time.

SBS envisaged that a company such as Sears Roebuck, with stores throughout the country, would lease on-premises earth stations, or terminals, and enough satellite capacity to interconnect them, from SBS. Customers could lease as much capacity as they needed for regular service, with the ability to lease extra capacity on demand. SBS bought six Ku-band satellites providing very high-quality performance. Such a large purchase of satellites is unheard of for a startup system. The order was very good for Hughes because it allowed them to start a new generation of satellites which has sold very well. The satellites and the sophisticated terminals built by IBM were the key components of an all-digital system. The digital feature meant that the system would be good at handling high-speed data, television, voice, facsimile, and electronic mail. One selling feature of the system was that it would reduce the need for many special services. As an example, it could carry all the inside mail at night when the system would be lightly loaded—the assumption being that the largest fraction of a typical multisite company's mail is sent to other employees in the same company.

SBS became operational in 1980 but was never able to get enough customers to achieve a satisfactory level of loading. Most of the capacity of SBS was used for handling voice, but while voice services are now being processed digitally, voice is not a large user of digital capacity, which SBS had to sell in abundance if it was to succeed. The lack of enough customers, and the slow growth of demand for high-speed data and other special services that would have justified the expensive digital ground stations, meant that the SBS system was greatly overdesigned. The extra satellites were also a heavy expense.

American Satellite Corporation (AmSat) provided a strong contrast to the SBS approach. If SBS was the Cadillac of the market, AmSat, a Fairchild affiliate now owned by ConTel (Continental Telephone), was probably content to be the Chevy. Whereas SBS's design philosophy was to offer its customers a very sophisticated terminal from the beginning, even before an immediate need had been established, AmSat used very inexpensive terminals and gave the customer only what he needed at the time. AmSat also saved on expenses by leasing capacity on other people's satellites.

SBS never became financially viable. IBM bought out its two partners and later traded the company and some of its satellites to MCI for shares of MCI stock, which have appreciated nicely. Several of its satellites have been leased by Hughes Galaxy, while others have gone to Comsat for Comsat Video Enterprises, a service to hotel rooms.

Next we will look at domestic satellite systems in countries other than the United States.

■ PALAPA

Palapa is the domestic system of Indonesia that also offers connections to a few of its regional neighbors and allies. The system was conceived by Indonesia as a way to tie together its farflung archipelago. Its population of 175 million people live on thousands of islands scattered over 5 million square miles of ocean. Linking all these islands together by any means other than satellite is a formidable problem indeed. Palapa's satellite beams cover the whole Indonesian archipelago, as well as Thailand in the west and the Philippines in the east. Palapa also includes a certain amount of service among countries of the Association of Southeast Asian Nations (ASEAN). The Indonesian case shows how an otherwise difficult communications problem can be solved simply and relatively cheaply by turning to satellites. Many of the islands have very limited needs that merit only the simplest ground equipment, even though the whole system is large enough to justify going first class.

The job of supplying 2 satellites, the central control station, and its 9 largest earth stations was contracted to Hughes Aircraft. The original concept called for about 40 ground stations, including those supplied by Hughes. Hughes did the complete design and delivered a turnkey system, with the two satellites having been put in orbit in 1976–77. Palapa operates in the C band.

To take account of satellite aging, Indonesia ordered a second generation of satellites from Hughes, called Palapa B. The first Palapa B was put into orbit in June 1983 after being launched on the space shuttle. The second Palapa B became historic. It and a Westar twin were launched on the space shuttle in February 1984 but were left in a strange and useless orbit when the perigee assist motors (PAMs) on each failed. (The failures of the rocket nozzles on the PAMs showed the danger of launching two identical satellites at the same time. If one is faulty, the other may contain the same fault. According to Comsat people who helped engineer the system, Palapa's managers were advised not to launch but they went ahead anyway.)

In November 1984 a shuttle crew successfully recovered both satellites and again made history. In a few years' time, the Westar satellite was refurbished and resold to a new owner. That satellite—now called Asiasat—was launched by the Great Wall Industry Corporation in Beijing on a Long March rocket in April 1990 and is controlled from a ground station in Hong Kong. Asiasat is a joint venture of Cable & Wireless of London, Hutchinson Whampoa of Hong Kong, and the Chinese International Trust and Investment Company of Beijing. Business has been so good that this consortium will order a second satellite even as its first one is being introduced. Asiasat's customers are Hutchinson Telecommunications and Capital Communications of Hong Kong, the government of Burma, the communications agency of Mongolia, and the Korean firm Data Communications Corporation. Other potential customers include Thailand, Nepal, Pakistan, and Bangladesh. Most of the satellite capacity will be used to broadcast TV throughout Southeast Asia.

Palapa B3, Indonesia's replacement for Palapa B2, was successfully launched in 1987 on a Delta rocket. In April 1990 the refurbished Palapa B2—now called Palapa BR2—was successfully orbited by a Delta rocket.

Because of the large investment Indonesia had made in ground equipment, its second generation satellites were in a C-band configuration like the first. Many countries have opted for Ku band, which has various advantages but does require new equipment on the ground.

■ MORELOS

In the early 1980s Mexico began to place large orders for domestic use of Intelsat transponders. At one time, they said they wanted to lease 40 transponders! Clearly Mexico had become very conscious of what satellites could do to bring the benefits of modern communications to their whole country. Mexico, like Canada, saw that satellites can play a special role in nation-building.

In 1982 the Mexican Secretariat of Communications and Transportation ordered two satellites from Hughes. They chose to have service on both C and Ku bands. Their satellites were launched on the space shuttle in June and November of 1985. In addition to the satellites, Hughes built a complete control station and delivered a turnkey system.

Figure 10. Technicians at Kennedy Space Center prepare to offload the Palapa (left) and Westar satellites retrieved by *Discovery*'s astronauts in November 1984. (Photograph courtesy of NASA.)

Mexico uses its satellites for educational and commercial TV, telephone, data services, and radio. The satellites permit Mexico to provide services to all areas of the country with good signal strength so that antennas can be small. Before satellites, Mexicans watched programs only from local TV stations; now they can receive programs that originate in all parts of the country.

▪ AUSSAT

Australia is an excellent example of how satellites can supply a host of communications services to a thinly populated nation at a reasonable cost. Australia may be the country that comes closest to satisfying the satellite designer's dream, in that it needs all kinds of service. Aussat is thus more versatile than almost any other satellite.

Under a turnkey systems contract, Hughes delivered three satellites for Aussat use, two control stations—one for Perth and one for Sydney—and other facilities. It also supplied launch and checkout services. The first two satellites were launched by the space shuttle in 1985. The third rode to orbit via Ariane in 1987. All operate at Ku band.

Aussat is used for communications between major cities, where it provides TV links for network TV, regular telephone and data services, and maritime and air traffic control services. It also serves the outback with communications, TV of such signal strength that small dishes will suffice, and other special services. The use of the satellites for air traffic control may be a first.

When the Australians recently decided to buy a second generation of satellites, they asked potential suppliers to bid on satellites delivered in orbit using Chinese Long March rockets. Hughes won the order on the assumption that an export license would be issued to allow the launches on Long March rockets. Then the United States placed a temporary hold on export of the satellites because of political unrest in China; the temporary hold was partially lifted in early 1990 to permit export of interface data such as the location of bolt holes, power requirements, and the like. In view of the fact that the U.S. government permitted the launch of Asiasat on Long March in early 1990, it is likely that Aussat will get permission for launch in 1992 when it is ready.

■ INSAT

About 10 years ago, after the United States completed the prescribed tests of ATS-6, an experimental geosynchronous satellite, it was made available to India for use in an educational experiment called SITE (Satellite Instructional Television Experiment). Thousands of Indian villages were supplied with TV receivers and makeshift satellite dishes. They received instruction on farm prices, economic news, farming methods, birth control, and other subjects considered valuable for improving their standard of living. The government found the results impressive, and SITE became a model for the development of a satellite system for India.

To increase their confidence that they were doing a good job in designing their domsat system, India asked Comsat to oversee the whole design process. Unlike many Third World countries, India has thousands of people trained in physics and electronics who can design a satellite system; even so, they asked Comsat to check their design and procedures to ensure that no critical points had been overlooked.

Insat, as India's domsat system is called, was originally planned as an educational system, but it has also become a successful commercial venture. Millions of TV receivers are now being built in India, as well as antennas. While some people in the United States have criticized India's departure from the educational rationale with which it began, once a communications system is in place it can be used for whatever purposes its owners decide are worthwhile. The commercial success of Insat is difficult to fault—especially if we consider that a good part of its capacity is still reserved for educational purposes. Insat's capacity is sometimes used for national broadcasts in Hindi, expressing a governmental desire to promote a single language for the country; at other times, it is used for local service in local dialects.

This treatment of domsats is not all-inclusive but should give a picture of how widespread satellite usage has become as more and more countries recognize the opportunities opened up to them by this new communications medium.

4

SPACECRAFT LAUNCHES

I f building communications satellites and operating them while in orbit is the largest commercial venture in space, the second largest is the business of getting satellites into orbit in the first place. Before the private satellite communications system was created, the government was the only launch agency in the United States—and therefore the only customer for launch vehicles, which it purchased from private suppliers. After the commercial satellite communications industry was established, space launches nevertheless continued to be under the control of the government, for financial as well as political and national security reasons.

In addition to taking into account NASA's expertise in launching rockets, the Bureau of the Budget (later to become the Office of Management and Budget) considered the selling of launch services to private companies such as Comsat, RCA, and Western Union as a way to help defray NASA's huge operating expenses. In the 1970s, when planning how much to charge for launching commercial satellites, NASA—under instructions from the General Accounting Office—set its prices high enough to cover all actual recurring costs and identifiable nonrecurring costs, but did not try to recover "sunk costs" having to do with research and development on launch rockets or launch facilities. This method of pricing became established policy early in the game. Later, under instructions from Congress, NASA reviewed its charging practices and discovered that commercial customers were not being billed for the costs incurred by the Air Force in assuring that an errant rocket would be destroyed before it could do any damage.

Having assured Congress that it was charging full price, NASA had to modify its practices and added on range safety costs.

Once development of the space shuttle was under way, the policy of government-supplied launches continued, the only difference being that once the shuttle was in operation, the government would have some choice of which type of vehicle to use—expendable launch vehicles (ELVs) purchased from commercial companies or the space shuttle. Then, in 1978, when the Carter Administration threatened to cancel the shuttle because of delays and cost overruns, the shuttle's protectors came up with a proposal that the government should close down every other launch program and force all payloads—commercial, civil, military, and foreign—to fly into space aboard the shuttle. Some of the creators of this "shuttle-only" policy used the argument that money would be saved by getting rid of the launch crews for the other rockets—the usual argument for a monopoly. But the real reason was the desire to ensure that the shuttle could not be canceled, once it had become the only game in town.

When setting a price for a launch on the shuttle, NASA did not attempt to recover even its recurring costs immediately, as it had done with ELVs, but rather set a price that ostensibly would cover all recurring costs once a certain launch rate was achieved. As the cost of the shuttle grew, all financial realism in pricing policy went out the window, and prices were set for purely political reasons.

Faced with the shuttle's artificially low prices and the Carter Administration's adoption of a shuttle-only policy, the Big Three launch vehicle manufacturers—General Dynamics, McDonnell Douglas, and Martin Marietta—silently, if regretfully, accepted the decision that the shuttle had become the U.S. flag carrier of the launch business. This meant no more sales of ELVs to NASA for launching commercial communications satellites, and no more contracts to launch NASA and NOAA payloads. The only exception to the shuttle-only policy was a provision that NASA and the military should fly out the remaining stockpile of Titan, Atlas, and Delta rockets, and they proceeded to do so.

■ AIR FORCE CELVS

The Air Force tried to convert as many of its payloads as possible to fly on the shuttle. In some cases, this was done at great cost, with the whole satellite configuration being changed to fit the contours of the shuttle payload bay. The Air Force did its best to follow the party line; it also made plans to launch the shuttle from its facilities in California, spending several billion dollars to modify a launch pad to accept the shuttle.

By the early eighties the Air Force was having second thoughts about the wisdom of the shuttle-only policy; recognizing that certain secret payloads would have to be launched even if the shuttle were grounded, the Air Force decided in 1984 to seek Congressional approval to develop the so-called CELV, or complementary expendable launch vehicle. NASA agreed to the Air Force's proposal to buy ten CELVs from commercial manufacturers, on condition that no more than two per year would be launched. Thus the Air Force was allowed to bend, but not break, the rule that the shuttle was the approved way to get into space. The CELVs were ordered, but unfortunately none were ready in time to cover for the shuttle when the *Challenger* accident occurred in 1986.

The Air Force's CELV order was won by Martin Marietta and took the name Titan IV. This is a much larger version of Titan than its predecessors, one capable of launching about twice the payload of the Titan III. The first Titan IV was successfully launched in June 1989. (This long lead time on CELVs is typical of complex space hardware; one usually has 4 or 5 years to wonder whether the right decision was made before there is evidence pro or con.) Meanwhile, the Air Force had also decided in 1984 to refurbish 15 Titan IIs for backup to the shuttle. The Titan IIs were ICBMs that had spent about 20 years in missile silos. A decision was made to phase them out of the missile force, and they were scheduled to be junked. Their refurbishment meant that the Air Force also had some backup for the smaller launchers that were being phased out. But like the Titan IV, the refurbished Titan II was not yet available when the *Challenger* was lost.

To add insult to injury, the stockpile of ELVs to which the Air Force turned in that critical moment failed as well. The last remaining ELVs each had a different failure diagnosed; there was no pattern, but at least two factors were at work. First, people who know they are going out of business right after the last launch are not as motivated to do a good job as they were when they thought they would continue in business; in fact, many people left the various ELV programs before those last, failed launches. A second factor was that the last rockets were the tail end of the stockpile and had gradually deteriorated with age.

So despite its precautions, the Air Force wound up being held hostage to the shuttle for about four years. The *Challenger* accident caused the Air Force to campaign even harder for "assured access to space."

■ ARIANE

The U.S. government's decision to force a shuttle-only policy on the Western world was challenged in 1980 by the arrival on the scene of Ariane. Back in 1974, the Europeans had developed a strong desire to create their own independent launch capability when NASA (at State Department insistence) turned down their request to launch their Symphonie satellites. The U.S. decision was a response to the initial unwillingness of the Europeans to agree not to compete with Intelsat. After a long hesitation, the Europeans promised to restrict the Symphonie satellites to experimental uses. NASA then launched two Symphonies, in 1974 and 1975. Nevertheless, the Europeans decided after that experience to develop their own launchers as rapidly as possible.

Once Ariane had flown successfully, customers around the world, including Intelsat, were no longer captive to the U.S. policy of launch-on-the-shuttle-or-be-grounded. Ariane broke the U.S. government's monopoly on space launches.

Several American companies had satellites they wanted to put into orbit and were willing to take a chance on the fledgling Ariane. Spotting the trend toward Ariane, proponents of the shuttle immediately cried foul and charged Arianespace with unfair competition. But when the U.S. Trade Representative's Office investigated the charges, it concluded that Ariane was no more heavily subsidized than the space shuttle.

Both the U.S. government and the Europeans continued to complain that the other side was subsidizing its launch industry, and both had a legitimate argument. When Europe was trying to become known as a reliable supplier of launch vehicles, it assured customers that Arianespace would always bid whatever price was necessary to win about 8 orders per year—their desired workload, and about 1/3 of the world market. They said they did not really want to corner the market. The United States, on the other hand, got into the habit of underpricing the space shuttle because we wanted to have it accepted as the primary way to get into space. During the five years from the first shuttle flight to the *Challenger* loss, shuttle prices were tied to costs only in a very nebulous way. We could force our own government's customers to ride the shuttle and could reach internal agreements on how much they should pay; but if we wanted commercial companies to buy rides on the shuttle, they had to be shown that shuttle prices were competitive. Ariane was the big competition.

Frederic d'Allest, chairman of Arianespace, once commented that it was a pity the United States did not raise shuttle prices so the Europeans could raise Ariane prices accordingly; then we would both stop losing so much money. What he was saying to Intelsat was somewhat different, however; he was always willing to bid a dollar lower to beat the competition for launching Intelsat's satellites.

In January of 1986, at the time of the *Challenger* accident, Ariane had a worse reliability record than the shuttle, and in May of that year suffered an outright launch failure. Nevertheless, Ariane began picking up orders for launches of American satellites by companies willing to take a risk. Through its marketing agency, Arianespace, Ariane eventually garnered half of the world's launch business, including many orders from the United States.

On its tenth anniversary on October 27, 1989, Ariane 4 launched Intelsat 6, the largest commercial communications satellite ever built. With great pride, Ariane now offers a full line of expendable rockets based on a very mature technology, and its current reliability record compares favorably with that of U.S. expendables. The U.S. percent of success is in the mid-90s compared with 86 percent for Ariane. The February 1990 failure of an Ariane rocket to reach orbit was the company's first launch disappointment since May 1986. After discovering the cause (a cloth left in a feed line), Arianespace reported that it would resume flights in August 1990 and would conduct 6 launches before year's end, counting the February failure. As of March 1990 Ariane had on the books 34 satellites remaining to be launched at a price of $2.4 billion, with 9 new contracts having been signed this year. Arianespace said the backlog would be flown out at a nominal rate of 9 per year.

■ COMMERCIALIZING THE U.S. LAUNCH INDUSTRY

The *Challenger* accident led to a complete reversal of U.S. policy regarding use of the shuttle. After what seemed an interminable period of paralysis, the White House announced in August 1986 that while it would still honor commitments to foreign governments to launch shuttle-unique payloads such as the European Space Agency's EURECA (European Retrievable Carrier) and Spacelab, commercial payloads would no longer be welcome on the shuttle; human life would not be put at risk for purely commercial purposes. According to the prevailing post-*Challenger* wisdom, only government missions had sufficient importance and priority to justify a shuttle flight. And even some government payloads were judged not to merit launch on the shuttle.

In parallel with the decision not to use the space shuttle for commercial launches, the White House announced a new commercial launch policy. Companies that wished to operate satellites would solicit bids for launches directly from rocket builders; they would not have to go through NASA as an intermediary. NASA was also told not to buy any more rockets to be launched under its own management. Rather, it was to buy launches on the commercial market, just as the private satellite companies were told to do.

NASA responded with a two-phased approach. Phase I was the purchase, on a noncompetitive basis, of about 9 launches for its immediate needs. Phase II involved a competition for small, medium, and large booster launches, wherein NASA asked for prices on a few launches, with options for larger quantities for later delivery. This process finally ran its course during 1989; NASA now buys launches competitively for large payloads but reserves the right to launch sounding rockets in its own way.

NASA's decision to go back to expendable rockets, while a positive development, was a total reversal of policy—not a small change. Under the old policy, Martin Marietta did not even dare approach potential customers for a ride on a Titan III for fear of antagonizing NASA, to which it was selling external tanks for the shuttle. The new commercial launch policy that totally reversed this situation had actually begun under the 1984 Commercial Space Launch Act. That year saw the *Congressional Record* bulging with the debate about what role the departments of Commerce and Transportation and NASA should play if we did in fact want to commercialize the launch business. As time went on, Transportation took on the major responsibility for developing policy and licensing launches of commercial payloads.

For example, in 1988 NASA and the Air Force, with the Department of Transportation acting as the one-stop shopping agent for the launch companies, completed arrangements to lease launch facilities to the launch companies. Reasonable rates have now been established whereby private companies can use government sites. Negotiating the terms on liability in the event of launch failure was not so easy. A launch vehicle going off course is potentially very destructive, even given range safety features that are supposed to destroy it before it can do any damage. Insurance to cover this liability is extremely important to launch suppliers. From the beginning, when NASA launched a payload for a company such as Comsat or RCA, it billed the companies for the cost of the launch service, even if the rocket failed. NASA left to the customers only the problem of worrying about insurance to cover replacing any lost satellite. That policy had both good and bad features, but it did protect the suppliers of expendable rockets from any liability to customers, since the government assumed liability for any damage due to a launcher going off course.

Since 1988 commercial launch companies have had to worry about liability—a factor that can weigh heavily on even a large company. There was quite a delay in settling the question of what fraction of the liability the individual companies would carry and at what point the government would assume liability. An arrangement was finally enacted into law stating that the private sector may be asked to assume the first $500 million of liability, although in practice the government has usually settled for more limited liability (in the range of $200 million). The U.S. government provides coverage from the agreed-upon level up to $1.5 billion. These ranges seem quite reasonable, since it is unlikely that damage will exceed $1.5 billion. This decision broke the logjam, allowing launch companies to get insurance—although at high prices—on whatever liability they must cover before the government takes over.

The commercialization of the launch industry received an enormous boost in 1986 when the Air Force initiated its Space Launch Recovery Program. The Air Force had started buying CELVs back in 1984 from commercial suppliers, two years before the *Challenger* accident, but this new program committed $10 billion to new launcher orders. The Air Force ordered 23 Titan IVs to launch Defense Support Program missile-warning satellites and other comparable payloads, and eventually increased the order to 41. It also ordered 20 Delta IIs to launch Navstar satellites (see Chapter 6) and 10 Atlas IIs to launch DSCS satellites. Since these orders are taking several years to fill, the Big

Three launch companies have felt confident that their production lines will be open for a long time, allowing them to respond competitively to bid requests from commercial customers.

Thus, following the abandonment of the shuttle-only policy, the Big Three should have been able to reenter the now officially commercialized launch market with great hopes of reaping good profits in a seller's market. There were many projects in the pipeline that had been awaiting a ride into orbit on the shuttle and that now had to turn elsewhere for launches. The launch companies had been given the green light to go out and capture this business. But the situation was confusing. The Big Three no longer had to worry about unfair competition from the shuttle, but they had no clear idea what the government would do about all the new foreign competition that had appeared on the scene in the interim. Not just the Europeans but even the Chinese and the Soviets had begun to tout their willingness to launch U.S. payloads into space, using their own rockets—the Chinese Long March and the Soviet Proton. The Big Three knew they had to compete on the open market among themselves to get orders from NASA, NOAA, and commercial customers, but the ground rules by which these foreign suppliers would operate were not at all clear.

The U.S. launch companies believed they had the right to expect some help from the U.S. government in controlling the game, or at least ensuring a level playing field. But the U.S. government was more concerned about the short-term problem of how to get certain payloads into space as quickly as possible than about how U.S. policy would affect the commercial launch business in the long run. Would the U.S. government help by establishing a policy on international launch competition, or were the U.S. companies out there on their own? If they were on their own, then one had to ask whether it made financial sense for a private company in the United States to invest money in a field of business where some of the competitors are sponsored heavily by foreign governments primarily interested in developing indigenous industry.

Both the Soviet and Chinese governments were rather frank in saying that they intended to quote prices below normal U.S. prices, leading critics to charge that they were "buying in." While it is very likely that neither country knows what their actual costs are, it is also quite possible that their real costs are less than ours. The Soviets have been building rockets for a long time with few design changes and must have benefited by long production runs and high launch rates. Ever since the shuttle came on the scene in 1981, the United States has built very few ELVs; consequently, our unit costs have continued to rise. These nonmarket-economy foreign agencies have said they are willing to modify prices to suit us (that is, to set "fair" prices), but their needs for hard currency will no doubt tempt them to continue their policy of low bidding.

But are the Europeans really a different type of competitor from the Chinese and Soviets? The Ariane rocket—the stiffest competition to U.S. launchers—is built by a consortium of private companies and government entities. The largest investor in Ariane is the French government. Some people have complained that Ariane's prices, like those of the Long March and Proton, are set far below actual costs for reasons of industrial policy—or that costs are not even known to Arianespace. Arianespace says that it is not in business to lose money; even if the French government is willing to sell launches below cost, the private companies involved want the venture to be profitable, they claim. That may be true in theory, but in fact the other investors in Ariane are companies and banks in which governments hold major stakes. Many members of the consortium are also members of consortia that build satellites to be launched by Ariane. In other words, in Europe significant government ownership of the aerospace industry is the rule, not the exception. In the United States the aerospace industry is also highly dependent on the government for business, but the government does not own major shares of aerospace companies. To the extent that a corporation is run by management, the two situations are essentially identical; to the extent that the shareholders run the company, the European situation is vastly different, in that the European shareholders—the governments—have industrial-policy considerations that must be taken into account.

My view is that Europe will base Ariane's prices on the desire to optimize the economics of the European space industry. I believe they consider this industry to be a vital component of their plan to stay on the leading edge of technology. If so, this is important to them all out of proportion to its size, since they see it as a major factor in their future economic competitiveness and business success.

The Big Three launch companies expect the U.S. government to offer them some kind of protection from foreign competition on launchers. All have made major investments that can only be amortized over a reasonably long period (although Martin Marietta chose to write off its investment more rapidly than expected and has shown signs of a waning interest in commercial orders). However, it remains to be seen whether they will enjoy a really stable policy. Of course the United States also wants to succeed in space commerce, as President Reagan announced in February 1988, and the government seems to recognize that a viable commercial launch business requires a steady stream of orders to keep production lines open. Orders from the Air Force and NASA have satisfied this need admirably for several years. In this respect, at least, the American launch companies have the advantage over the European competition; no U.S. Air Force orders are going to Ariane. (We seem to know very little about European military orders going to those companies that build Ariane rockets.) Our European friends might consider their inability to bid on U.S. government launches as unfair; but do they let us bid on launching their SPOT and DBS satellites? Obviously not.

■ COMPETING FOREIGN INTERESTS

When the Australians asked for bids on second-generation Aussat communications satellites, they asked bidders to include the cost of launches in the price and to include the Chinese as potential launch suppliers. The Australians chose the Hughes bid for Hughes satellites and Long March rockets (priced at about half what U.S. suppliers would have charged) to put them in orbit. After a rather lengthy debate, the U.S. government approved the Hughes proposal to Aussat and a second Hughes proposal to Asiasat.

In the discussions inside the government about whether to allow the Chinese to launch American-built satellites, everyone recognized that Air Force orders would keep the major U.S. launch companies in business for quite some time, so there was relatively little concern about protecting their competitive edge as commercial suppliers. But concern about the transfer of advanced technology in communications payloads was certainly a primary consideration in the negotiations. The Carlucci mission to China which settled the agreement on how many American satellites could fly on Long March rockets had to take into account a whole host of foreign-policy considerations; it is fair to say that the outcome (9 satellites over a 6-year period, with prices to be held close to market economy prices) was meant to trade certain advantages to the Chinese for important American objectives—including giving China incentives to sell to us rather than to certain other nations.

Asiasat, which has already flown on a Long March rocket, represents 15-year-old technology, so we will not know until Aussat comes due in a few years whether the administration is serious about its policy to permit Long March to launch only old designs and to restrict the latest communications technology to flights on U.S. launchers. Congress reinforced the restriction on what level of technology to permit on Long March as recently as December 1989. This policy tends to give U.S. launch suppliers a captive market—albeit one of limited size. It also helps ensure that Chinese rockets will be available if U.S. rockets are grounded again, as they were in 1986.

The events of June 1989 in Tiananmen Square caused our Chinese policy to harden. Many in Congress said that we should not honor the Hughes agreement in view of human rights violations by the Chinese government. President Bush has said that he does not want to throw away the progress made on improving relations and trade with China, but the Chinese government's response to students protesting the lack of democracy caused us to put the sale of munitions on temporary hold. Satellites are considered munitions for arms control purposes, so the Chinese launch of American satellites also went on hold. As of this writing, the administration seems much more inclined than Congress to assure the Chinese and Australians that they can count on our shipping satellites in time for their launch a few years hence.

Australia was heard from a second time in mid-1989. The *Washington Times* of July 29, 1989, reported that a spaceport is being planned for the northern tip of Australia (Cape York), and a number of American companies will be participating in the project. According to the report, a United Technologies subsidiary, USBI, will assist in integrating payloads on the Soviet rocket Zenit. This would give Australia the ability to launch satellites into geosynchronous orbit from a site 1,000 miles closer to the equator than Cape Canaveral. (Ariane launches even closer to the equator and thereby achieves about a 10 percent advantage over the U.S. site.) The idea of course is to make launches on this site so inexpensive and attractive that U.S. satellite companies cannot ignore it. An arrangement using such a launch site would split the world market for launches into still more splinters—making commercial success for American launch companies more doubtful.

Will the U.S. government find it in our national interest to let the Soviets launch a few U.S.-built satellites? Certainly we had more reasons to want to cut deals with the Russians than with the Chinese, even before the latest developments in Eastern Europe. Although there is now an absolute prohibition on letting the Soviets into the U.S. launch market, we know that there will be continuing pressure to increase trade of all sorts between our two nations, with the launch business likely to be part of the overall package of items up for discussion. If the United States decides to take extreme measures to assist in bolstering Gorbachev's staggering economy, then Soviet launches could be a big-ticket item in enabling the Soviets to earn hard currency with which to buy American products.

Past practice suggests that we will find innumerable reasons to open the U.S. market to outside suppliers, even though those who want access to our markets seldom believe it makes sense for them to share part of their market with us. The Europeans have placed three orders in the United States recently: one launch of a Eutelsat satellite, one BSB satellite, and one Italian. By contrast, Arianespace has garnered dozens of orders for Ariane launches from U.S. companies. (To be fair, we should include Intelsat and Inmarsat launches; these have typically been split between the United States and Europe.) The Chinese and Russians have no commercial satellites for us to bid on launching, and they will not ask us to bid on launching their government's spacecraft. Dealing with them is not a two-way street; the question is whether we see some larger advantage in having them launch U.S.-built satellites.

To some extent, our poor record in negotiating favorable trade agreements on space launches is the price of leadership. We still carry the principal burden of maintaining good international relations in our various alliances, as the leader of the West. Because of these alliances, we are expected to act the part of senior statesman. We must take account of all the ramifications in terms of foreign policy and the cohesion of the alliances. Most of the people we trade with have more limited interests at stake and are much more willing than we are to engage in hard bargaining.

Nevertheless, in my opinion, U.S. government policy should distinguish between launching U.S. satellites for use in this country and

Figure 11. The first commercial Titan poised at Cape Canaveral Air Force Station awaiting liftoff in December 1989. It carried a Japanese commercial communications satellite and a Skynet communications satellite for the British Defense Ministry. (Photograph courtesy of the U.S. Air Force.)

launching satellites for the use of foreign entities. In the latter case, foreign entities certainly have the right to decide how their satellites will be launched; we can then decide whether there is some reason not to allow U.S. companies to bid on such orders. But it is something else again to permit foreign launch suppliers to launch U.S. satellites for U.S. companies to use here at home, and I believe we should control who is allowed to bid in accordance with some kind of understanding of the ground rules for the competition. We should not sign up for a one-way contest. Critics will say that we cannot be inconsistent and ask the Japanese to open their markets while we close ours to other launch suppliers. The point is, all these issues involve questions of bilateral balances; since Japan enjoys a $60 billion per year advantage over us in trade, we have every right to ask them to open their markets—especially when even key Japanese spokesmen admit their markets are rigged against us.

Senator Albert Gore, chairman of the Senate's Science, Space and Technology subcommittee, believes that foreign governments make decisions on how to price items like launchers under broad government policy, and stated that we would be naive to assume that the normal rules of competition apply. We must have intergovernmental agreements on how the market is to be shared between us and our friends and competitors. He said we should not assume that the free market will settle such matters properly, since there is no free market in launches. It is a manipulated market, in his view, and we have to play by the same rules as our competitors.

The launch business demonstrates that there can be no commercial success unless the government takes the necessary initiatives to create an international environment where U.S. companies can compete fairly. We should not pretend that our government does not help our own launch companies, but ordering government-needed rockets from them is only part of the story. If, when they go to sell a launch to a private customer or foreign government, their competititors are governments themselves, there may be no way to compete effectively. The U.S. government must reach agreement with our trading partners which precludes predatory pricing; if that is done, our companies should be able to compete. If not, then there may be no way to sustain a commercial launch business in the United States.

It is cause for optimism in this regard that the Vice President has become personally involved in the issue and has spoken out about unfair competition in the launch business. He has met our European competitors to discuss the launch business; those European partners are now pressing the United States to take a firmer stand about additional Chinese launches and possible Russian launches. A new U.S. position on foreign launch of American satellites was supposed to be ready by mid-1990 but was deferred pending further discussions with those suppliers.

We should not focus so much attention on the international scene that we ignore the need for American companies to learn a few tricks of their own. U.S. commercial launch companies must do more to reduce costs. If the government protects them totally against foreign competition, the result may be that they remain sluggish and uncompetitive. What we need is enough competition to keep them on their toes, to make them see one another as competitors and not as fellow eaters at the government trough. If they are able to reduce costs, they will, in the process, help enable new space ventures to succeed. Many such businesses see the high cost of getting into space as a major deterrent to their success. It would be ironic if we kept prices up to help the launch companies stay in business, and in the process discouraged various new ventures from progressing because of high launch costs.

▪ VIABILITY OF THE COMMERCIAL LAUNCH INDUSTRY

In view of all these considerations, can we assume that there will be a commercial launch industry in this country? Well, there already is. All of the Big Three have booked orders for commercial launches. They believe there is a market for about a dozen commercial launches per year of U.S. payloads, and they expect to capture their share of the market. Other payloads on the world market need to be launched; they expect to capture some of those orders as well. As of early 1990, they apparently have about 20 orders for commercial launches through a combination of domestic and foreign sales (although some may be double-booked). When I recently asked a key launch company executive how many commercial orders he had, he said the answer depended on what I meant by commercial; now that NASA and NOAA have been told to buy launches like anyone else, he is trying to get used to the idea of adding in their orders as well.

McDonnell Douglas, maker of the Delta, had a backlog of 32 orders as of September 1990. The nonrecurring cost of their new Delta II rocket—their entry in the commercial launch business—is underwritten by an order for 20 launches of Navstar satellites—3,200-pound satellites which Delta places in a circular orbit about 12,000 miles high. Navstar is an advanced navigation system of extreme accuracy (see Chapter 6). As of September 1990, 8 Navstar satellites have flown.

Soon Delta will be upgraded to lift 4,000 pounds to a 12,000-mile orbit. Delta contains a large proportion of old technology, but the avionics have been modernized and simplified, saving money in the process. Delta, including its ancestor the Thor missile, has a total of about 500 launches behind it. The second stage has never failed to operate properly. The third stage is a Payload Assist Module (PAM), which is also used as an upper stage for the shuttle. In that mode it has launched a number of satellites into geosynchronous orbit from the shuttle. There is an obvious advantage in having a common PAM for both the shuttle and Delta, but PAMs come in three sizes, each with different weight-carrying capability and perhaps different failure modes. The PAM had an enviable launch record until the double loss on the shuttle in 1984 of both a Westar and a Palapa satellite (see Chapter 3). The accident was the result of introducing a new light-weight rocket nozzle on PAM—a supposedly "better" nozzle. Even so, PAM's record is excellent—43 successes out of 45 attempts, or 95 percent.

The Delta rocket's first commercial launch was in August 1989, when it put into orbit the first satellite of British Satellite Broadcasting (BSB-1). A year later, it launched BSB-2. Counting Air Force and commercial orders, McDonnell Douglas expects to be able to sustain a launch rate of about 10 per year. In 1989 the Delta achieved 8 launches—5 for Navstar, 1 for the Strategic Defense Initiative (Star Wars), 1 for NASA's Cosmic Background Explorer (COBE) satellite, and 1 commercial. In 1990, it expects 12 launches (8 completed as of September 1, 1990). The Delta is the smallest and least expensive of the Big Three launchers. In the last 13 years, it has had 63 successes out of 64 attempts—a 98 percent record.

The Atlas II launcher, which was selected by the Air Force in a medium-launch-vehicle (MLV) competition a few years ago, is slated to launch 10 Defense Satellite Communications System (DSCS 3) satellites. It has been ordered for a number of other uses—10 for Navy UHF and EHF communications satellites built by Hughes, 1 for Eutelsat, 3 for Intelsat (2 Intelsat 7s and 1 Intelsat K), 2 for Hughes Galaxy domsats, and a number of NASA scientific satellites and NOAA GOES weather satellites. Atlas II's first commercial launch is slated for late 1990. As of September 1990, General Dynamics boasted 40 total orders for Atlas II—17 for Air Force, 6 for NASA, and 17 for private organizations. Atlas also has a good launch record, although not quite as impressive as Delta's.

The Martin Marietta situation is more complex. The company has a long and successful history as a launch supplier but mainly with the Air Force—not with NASA. In fact, for the past decade, its largest NASA launch contract has been as a supplier of the shuttle's expendable external tanks. When NASA decided that the shuttle was the wave of the future for launching satellites, Martin Marietta went along; after all, it was getting good orders for the external tanks. Then when the Air Force placed its CELV order, Martin Marietta got a billion-dollar order for 10 Titan IVs—an order which eventually was raised to 41. Although it has received a few orders for commercial launches, it has not been as aggressive as McDonnell Douglas and General Dynamics. After its attempt to launch an Intelsat 6 satellite in March 1990 ended in failure, it filed a suit in July in what some called a preemptive strike against its customer. The *Washington Post* reported on September 5, 1990, that Intelsat had struck back and filed suit against Martin Marietta for negligence. Commenting on the earlier suit by Martin Marietta, *Via Satellite* magazine editorialized in its September issue that "it took more than a little coaxing to get Martin Marietta to admit that its wiring error caused the mishap. Now it adds insult to injury." It seems that this series of events has further reinforced the impression that Martin Marietta is not looking to its commercial orders for any significant amount of business, even though it may very well bid on a few orders in the future—provided it finds conditions right.

▪ SMALL COMPANIES IN THE LAUNCH BUSINESS

A number of companies other than the Big Three are either in business or trying to get in business to offer special-purpose launchers (or launches). One of the largest costs of launching satellites is keeping the launch crews and facilities alive between launches. Most launch companies in the United States have low launch rates, with crews and facilities greatly underutilized. More than half the total cost of U.S. launches is attributable to this factor. All things being equal, splitting up payloads into smaller pieces and increasing launch rates would tend to alleviate this problem, and that is where the smaller rockets come in. But of course many other factors must be taken into account in sizing launchers to achieve optimal costs.

There has been a great deal of speculation about the size of the market for smallsats, lightsats, or, as some have called them, cheap-sats. It seems that there is a market, but its size is very uncertain. LTV Corporation is a large aerospace company with a small rocket called Scout that, in the last 20 years, has launched more than a hundred payloads weighing less than 1,000 pounds. It has had a very good track record. NASA has been the largest Scout user but has not ordered any Scouts lately. NASA recently began a competition for about 10 launches of payloads in the 250–500 pound class, many of which could be launched by Scout. LTV proposes to upgrade the Scout by using some strap-on motors made by a Fiat subsidiary. The NASA competition should be resolved early in 1991. Since the Scout inventory is almost depleted, LTV has reportedly tried to arrange a way to reopen production of the Scout, but the picture is unclear. While the Scout rocket can launch small satellites, it does not compete with the Big Three companies for communications satellites in geosynchronous orbit.

One of the bidders in the NASA competition is Orbital Sciences Corporation, a relatively new company offering small rockets. OSC has three products on the market. Its first, a transfer-orbit stage (TOS), is an upper stage for the shuttle, competitive with the PAM previously mentioned. OSC began development of TOS before the *Challenger* loss and has received two orders from NASA—one for the launch of the Mars *Observer* and one for the launch of the Advanced Communications Technology Satellite, both scheduled for 1992.

The decision not to use the shuttle for commercial launches greatly reduced OSC's opportunities for getting a significant number of orders for TOS, so OSC decided to enter a totally new market. It developed Pegasus, a launcher for small (600–900 pound) spacecraft. The first stage of their Pegasus launcher is an airplane, which can be flown to any desired launch point, preferably over water, and positioned to launch Pegasus in any direction.

OSC has borne all the cost and risk of developing Pegasus, recovering only a small fraction of the development cost each time it sells a launch to a customer. To finance this venture, OSC sold 20 percent of its common stock to Hercules, the company building the Pegasus rocket motors. Several years earlier, when it was just getting started, OSC had used a similar self-financing scheme to develop TOS. The money for TOS development was raised by setting up a research and development partnership, an arrangement that attracted much venture capital before the 1986 tax law changes made such things less appealing to investors.

Figure 12. The small launcher Pegasus is carried under the wing of an airplane, which flies the spacecraft to the desired location and positions it to launch in any direction. (Photograph courtesy of Orbital Sciences Corporation.)

In early 1990, having depleted its capital too far for comfort, OSC decided to "go public," that is, to put its shares on the market to raise additional capital. Plans were drawn for the announcement to take place on March 20, 1990. That day, the *Wall Street Journal* ran a front-page story to the effect that "three space nuts" who ran OSC were most likely going to fail in their attempt to launch Pegasus. The stock was quickly withdrawn from the market. Two weeks later, OSC succeeded in launching the rocket, which carried two small satellites into orbit. Two weeks after that, OSC went public successfully at a price higher than the price they had in mind originally. The *Wall Street Journal* carried that story on page 18. (Why the likelihood of failure was a page-1 story and the actual success was fit only for page 18 is one of those mysteries known only to the reporter and editor.)

In mid-1989 the Defense Advanced Research Projects Agency (DARPA) conducted a competition for small payload launchers on the theory that there will be a good market for small satellites and that small satellites, in turn, will create a market for small launchers. OSC won the order to launch a 1,000-pound payload. OSC will use a modified Pegasus, replacing the first stage (the airplane) with a Peacekeeper stage and launching the rocket from the ground. This ground-launched version of Pegasus is called Taurus. All together, as of mid-1990, OSC had contracts to launch three payloads on Pegasus for DARPA.

Space News of March 19, 1990, reported DARPA's intention to place 20 or 30 orders with various companies for additional small satellites to be used in weather and ground sensing, tactical communications, and battle-area monitoring. Thus DARPA's contention that there is a market for small satellites may well come true because of its own efforts to stimulate that market.

The success of OSC's products—like that of all companies in the launch business—depends on having a stable environment long enough to recoup development costs. OSC's experience on TOS was not encouraging; the shuttle policy did a 180-degree turn while OSC was in the middle of its development program, throwing all the company's sales projections out the window. Pegasus seems to be the best-established entry in the small-launcher market at this time, but since OSC does not yet have enough orders to cover development costs, its future is still unknown.

American Rocket (AmRoc) is another startup company that had plans to undercut costs of the Big Three by a large factor. Its founders believed that their hybrid engine concept would produce rockets at drastically lower costs than the prevailing technology. A hybrid engine combines a solid fuel with liquid oxydizer and, according to AmRoc, achieves a lower cost and a much higher level of safety over liquid-liquid rockets.

After plans for raising capital collapsed during the October 1987 stock market debacle, AmRoc spent a long time recovering to the point of conducting many static firing tests and completing construction of a test rocket. Unfortunately, their president, George Koopman, was killed in a car accident in the summer of 1989, and their launch in September 1989 was a failure. While they were reported to have orders for Strategic Defense Initiative (SDI) launches, the latest news is that they have decided to forego making and selling complete boosters but will sell only hybrid rocket motors to those already in the launch business. Certainly the jury is still out on whether they are a reliable entrant into the market.

Another start-up company, Space Systems, Inc. (SSI), has been developing and offering for sale a small expendable rocket called Conestoga. It is said to be able to launch 800 pounds into low earth orbit, but SSI has never hooked a customer and never completed a launch. SSI has launched sounding rockets, which have perhaps helped its credibility as a supplier of satellite launchers. In March 1989 SSI launched the sounding rocket Starfire 1 from White Sands; it carried a 600-pound payload for scientists at the University of Alabama who wanted to do experiments in very low gravity (see Chapter 8). The rocket rose 200 miles into space and gave them about 8 minutes of freefall microgravity. SSI had a "partially successful" second launch in November 1989, also for the University of Alabama. It was successful only in the sense that the payload was not destroyed when a gyro caused the rocket to fail early in its proposed trajectory. In May 1990 Starfire 3 flew successfully, carrying the 1,000-pound payload from the aborted Starfire 2 flight. The rocket again reached 200 miles altitude in a 15-minute flight which provided about 8 minutes of microgravity. While SSI has not done badly in orders for sounding rocket launches, this kind of business may or may not advance the cause of their becoming a supplier of orbital launches. In mid-summer 1990, SSI announced that it had run out of money and was laying off staff, pending finding another source of financing. Its financial backers said they could not justify putting any larger fraction of their assets into a space company.

Conatec, Inc., is another competitor in the sounding-rocket business, but its future is unknown at this time. E Prime is a small company that has agreements allowing it to use Peacekeeper components to bid on launches. So far it has not launched a satellite. Pacific American reportedly has some money for development of a low-cost ELV for small payloads in low orbit. Another small recent entrant is Microsatellite Launch Company, about which little is known at this time. Space Data has also been a launcher of sounding rockets; in 1989 it was bought by OSC. Space Vector is another small company in the sounding-rocket business which could be in position to bid on small satellite launches.

Another possible way to put a few payloads into space is by using the space shuttle's external tank. Normally the external tank, which carries liquid hydrogen and oxygen in two huge tanks joined together by what is called the intertank, is boosted to about 99 percent of orbital velocity before it is cut loose from the shuttle orbiter and allowed to crash into the ocean (mostly it burns up on reentry into the atmosphere). It would not be very expensive to boost those tanks all the way to orbital velocity—provided there was a small rocket engine on board. If put in orbit, they could be very useful as experimental laboratories (see Chapter 7) or as storage facilities for fuel, items awaiting repair, spare parts, trash, and other such materials.

If someone just wants to get an experiment into space for a few minutes but does not want it to go into orbit, the project could possibly be accommodated in the intertank. Certain experiments on microgravity, or density and composition of the upper atmosphere, could well be done on what is called the suborbital mode of the tank. On shuttle missions in which the tank eventually drops into the Pacific Ocean, there is a period of about an hour of microgravity before the tank reenters the atmosphere; on missions that call for dropping the tanks into the Indian Ocean, there are about 35 minutes of microgravity. These time periods are much longer than those provided by sounding rockets, but unfortunately the payload in the tank cannot be recovered, as it can from a sounding rocket. But if data from experiments in the tank could be telemetered down or even ejected in a capsule from the tank for parachute recovery, perhaps a few specialized customers might find this option attractive.

During the Reagan years, some members of the Economic Policy Council were vehemently opposed to any kind of subsidy for startup commercial space projects. The Bush Administration reiterated most of the Reagan space policy statements including the ''no-direct-subsidy'' clause. Fortunately, the Big Three did not and do not need subsidies: they need continuity to keep their production lines open. Air Force and NASA orders now on the books do this job very well. But the startup companies are another matter.

In my view, the Reagan Administration's no-direct-subsidy policy reduced greatly the odds for fledgling businesses to mature to the point of success. While the best of them do not want or need a subsidy as such, they do need someone to express enough confidence in them to enable them to get financing to fulfill the small government contracts they are likely to attract. Often the real dynamo to get such businesses off the ground is one or two small government orders that give a company credibility. Once the ball is rolling, many of these startup companies find ways to keep going and growing, even though they may have been only marginally viable at the beginning because of lack of sufficient capital. We should not forget that many of today's successful aerospace companies got their start when the government chose to buy their innovative products to fulfill its needs; the orders also enabled the creation of new enterprises. Air Products is an example of a $5 billion business which got its start selling liquid oxygen to the government; now its government sales are only a small fraction of total sales. Some of these new companies have the potential to return in taxes many times what the government may invest to get them started. DARPA has a good record of trailblazing in this field; what it has done on small launchers and small satellites illustrates how a few government orders can be a great stimulus to innovative thinking. It is interesting to see that those on Capitol Hill wanting to increase U.S. competitiveness are looking at the DARPA example as a possible prototype for a comparable civil agency. Such an agency in the Department of Commerce might sign contracts with new companies that would help them get started. The theory is that once their success has been demonstrated, the companies should be able to make it on their own.

5

REMOTE SENSING

The terms "remote sensing" and "earth observations" are frequently used synonymously to mean collecting information about our world in the form of radiation received at some remote point. Various forms of sensors can be mounted on platforms such as balloons, airplanes, or satellites to receive visible light, infrared (heat) radiation, or radar reflections. Remote sensing, while an old art form, has been considerably refined in the last few decades so that we can now use it to determine with great accuracy many important characteristics of the earth, its surface, atmosphere, and clouds; we can study runoff from snowpacks and rainstorms, the state of water in lakes, growth of plankton in the oceans, and many other parameters.

Remote sensing has a long history. People took pictures of the ground from balloons more than a century ago. The photographer George Lawrence took pictures from cameras suspended on kites and balloons and became famous by using a huge camera lifted by balloon to take pictures of the devastation caused by the San Francisco earthquake and fire of 1906. Balloons were supplemented by rockets in the early days of the century, and both were made obsolete by the airplane, the ideal platform for most photographic purposes. Aerial reconnaissance played an important role in World Wars I and II. At first pilots merely reported what they could see visually, but later they used handheld cameras to take photographs that were then analyzed by experts on the ground.

After World War II there was renewed interest in rocket photography because rockets could lift cameras several hundred miles high—a feat no airplane could aspire to. This gave new impetus to space pho-

Figure 13. TIROS 1 is given a vibration test at RCA in April 1960. The television cameras on this weather satellite were designed to take pictures of the Earth's cloud cover and transmit them to ground stations. (Photograph courtesy of NASA.)

tography and led to designs for satellites that would carry cameras and other sensors. The Air Force sponsored a number of studies which covered various uses for space platforms, including reconnaissance, mapping and geodesy, arms control verification, weather forecasting, and many others.

Civil remote sensing from satellites began in earnest in April 1960 when NASA launched the first low-altitude weather satellite, TIROS (Television and Infrared Observation Satellite). Based on early Army work, the first TIROS weather satellite flew at an altitude of about 500 miles and an inclination of about 48 degrees to the equator, giving it coverage of the United States and the tropics but missing important weather developments in the northern latitudes, including polar regions. In 1965 TIROS began flying polar orbits, thus covering the whole globe. At about that time, NOAA's predecessor, the Environmental Science Service Administration (ESSA), took over what were called operational weather satellites, and NASA continued with research and development of satellites such as the Nimbus series.

Several new generations of weather satellites have succeeded the first TIROS and continue to be operated by NOAA today. The successor satellites still fly at about 500 miles altitude and 98 degrees inclination—the so-called sun-synchronous polar orbit. The 98-degree orbit precesses eastward about 1 degree per day—just the right rate so that the satellite's plane of orbit is the same relative to the sun, regardless of the season. Therefore the satellite passes over the same part of the earth at the same time each day, year-round. The TIROS orbits allow the satellites to look at a swath of the earth about 2,000 miles wide every 100 minutes. For a number of years these satellites have included sensors that receive infrared as well as visible light and so give useful data day and night.

The Geostationary Operational Environmental Satellite (GOES) system joined its low-altitude counterparts in 1974 to give us another way of observing the earth. GOES flies in a geosynchronous orbit (22,300 miles high) and stares at the same piece of geography all the time. Because other countries fly geosynchronous satellites similar to GOES, the world meteorological community has continuous and essentially worldwide access to high-altitude data. The World Meteorological Organization, an agency of the U.N., coordinates collection and distribution of the data on a worldwide basis.

Both types of orbits—sun-synchronous and geosynchronous—have advantages of their own, and the two sets of satellites provide much more information than either one can do alone. From them we can learn about the atmosphere's cloud cover, moisture content, temperature, wind patterns, hurricanes, and so on. Information derived from them is used regularly for weather reporting and prediction. Most of us think we understand the weather better by seeing animated daily pictures of clouds from GOES satellites on the evening TV news.

▪ LANDSAT

Low-altitude weather satellites have enough resolution for studying the sky (about 1 to 4 miles) but are not adequate for studying details on the ground. For that purpose, civil remote sensing with satellites capable of relatively high resolution began in 1972 when NASA launched the first Landsat (called Earth Resources Technology Satellite, ERTS, at the time). Landsat, which flies in essentially the same orbit as the polar weather satellites, differs from them in that Landsat's much higher resolution allows detailed analysis of the surface of the earth. While TIROS wants to see clouds, Landsat wants to avoid them, looking at ground and water areas not obscured by clouds.

Landsat 1 was followed by Landsats 2, 3, 4, and 5. As of late 1990, Landsat 4, launched in 1982, and Landsat 5, launched in 1984, were still in operation. Landsat 6, now under construction, is expected to fly in late 1991. Over the years Landsat's images have been upgraded both in spatial resolution and spectral distinctions. Landsat's sensors detect both visible light and energy outside the visible spectrum. By this method, Landsat can make false-color images, revealing much information that is not available from visible light alone. The resulting images are invaluable for studying crops during the growing season and for identifying such problems as blight, damage from acid rain, and pollution in coastal waters. Landsat lets us look at crops worldwide, so we can predict whether the world's annual harvest will be adequate to feed everyone or whether we will have to draw down supplies from storage to tide us over until the next growing season. In addition to allowing us to study crops and forests, Landsat data can be very useful in the study of rock formations and minerals, snow cover, flooding, land use, and urban sprawl. Companies searching for mineral and oil deposits have become some of the most avid users of Landsat data.

To serve the needs of many individuals and organizations, there is a large industry in the United States and a few other countries that extracts from the great mass of Landsat raw data the specific information that meets their customers' needs. Although estimates of the size of this value-added industry vary, most observers agree it is in the range of $100–200 million per year. A good fraction of that amount represents purchases of data of one kind or another by federal, state, and local government agencies. Many local governments—and land-development companies, too—are interested in what is happening in the suburbs, such as subdividing of property, urban sprawl, and installation of water and sewage lines, electricity, cable TV, and telephone lines. Agribusinesses and timber and pulpwood managers can all understand their resources better by using Landsat products. The growth of this value-added industry shows how space-derived data can lead to new business opportunities of various sorts. Landsat's total sales are only about $25 million per year, but the value-added industry based on Landsat is larger by a factor of five or ten.

▪ SPOT

Many Landsat customers have welcomed a new entry into the remote-sensing field, France's SPOT (Système Pour l'Observation de la Terre) satellite. After all, most people believe they are better off when they have more than one place to buy what they need. The launch of SPOT broke the United States' monopoly on land-surface-sensing satellites, just as Ariane broke the U.S. monopoly on launchers. SPOT first flew in February 1986—fourteen years after Landsat's first flight—so it is not surprising that SPOT has features not found on Landsat, including stereo capability, ability to revisit areas more frequently, and higher spatial resolution.

SPOT itself is flown by CNES, the French space agency comparable to NASA, but a commercial company called SPOT Image, S.A., has the job of marketing the data. This French company, called SISA for short, owns a U.S. subsidiary named SPOT Image Corporation, or SICORP. SISA and SICORP sell pictures and other data collected by sensors of higher resolution than Landsat—10 meters for SPOT versus 30 meters for Landsat. SPOT II joined the earlier SPOT (now renamed SPOT I) in orbit four years after the first SPOT launch.

In 1987 the Soviet mapping organization, Soyuzkarta, made something of a stir by offering pictures of 5-meter resolution. But SPOT, like Landsat, offers digital data tapes as well as pictures; the Soviets offer only prints made from photographic film returned from orbit in capsules. Because the Soviets have not yet fielded an effective marketing organization, they have had only a limited impact in the remote sensing business. This situation may be about to change; at the Space Commerce conference in Montreux in March 1990, Charles Williams, president of EOSAT (see below), said that his company and Soyuz-karta were discussing an arrangement whereby EOSAT would market Soviet data. This would not only give the Soviets a commercial outlet but would give EOSAT a new set of data products to sell—probably a good business deal all around.

A second Soviet remote sensing initiative made the news in February 1990. *Space News* reported that the Soviets are taking orders for radar images made with their Almaz satellite to be launched in late 1990. The Almaz prototype satellite has been in operation for two years, with apparent success.

While the Soviets' impact is unclear, SPOT Image, by contrast, has become a major competitor with Landsat, affecting in complex ways the United States' efforts to turn Landsat into a commercial enterprise.

■ COMMERCIALIZATION OF LANDSAT

Because of the obvious importance of the data collected by satellite and the economic value of what we collect, remote sensing from space is frequently cited as an example of a field ripe for commercial development. Accordingly, the U.S. government decided in 1984 to commercialize the Landsat remote-sensing system.

During the Carter Administration, a study of Landsat's future was made, and in November 1979 a presidential directive was issued that called for transfer of Landsat to NOAA, as the weather satellites had been. NASA would keep responsibility for improving the performance of both these systems (as well as collecting scientific data), but it would not oversee routine data-gathering operations. The actual transfer to NOAA took place in 1983. The directive also stated that eventually Landsat would be run as a commercial business.

While there are valid arguments for giving responsibility to NOAA, unfortunately we have paid a price for moving satellites from NASA to NOAA management. In the first place, because its budget has typically been much more constrained than NASA's, NOAA has been reluctant to take responsibility for new satellites. For example, the proposal to have NOAA share the cost of the NASA Nimbus program (a weather-sensing system that is more advanced technologically than TIROS) was greeted with alarm, and NOAA continued to operate the older TIROS satellites instead (although some Nimbus-developed subsystems saw use on TIROS satellites). The transfer of Landsat to NOAA raised additional red flags at the Office of Management and Budget because of the large increase in NOAA's budget that this transfer would make necessary.

From the beginning the U.S. government has had uneasy relations with the Landsat project. In an earlier attempt to hold down the expense of the Landsat program, the OMB prohibited NASA from developing a real-time capability to process Landsat data. With such a capability, Landsat could be directed to cover trouble spots such as floods, fires, and famine and give rapid readout of results. If this had been done, Landsat would have been a much more valuable asset and would have attracted a much higher level of support in several government agencies than has been the case.

Since the mid-1970s there has been a continuing attempt by the OMB not only to hold down NOAA's budget but to get Landsat out of the federal budget altogether. Legislation intending to commercialize the Landsat system was finally passed in 1984. The term "commercialize" referred to the eventual outcome; right now the government is still incurring costs to operate Landsat. The plan was that when certain milestones were passed, a private company would operate Landsat with the goal of making the program profitable. A long-term contract was contemplated so that the company could justify investing its own money and have a fair chance of recouping any such investment with a reasonable profit. The Department of Commerce (DOC) initiated a process to create such a contract.

That process included holding a competition and selecting a winner which would operate Landsats 4 and 5 and would construct and operate the successor satellites, Landsats 6 and 7. From several bidders the DOC selected a company called the Earth Observation Satellite Company (EOSAT), a joint venture of Hughes Aircraft and RCA. The contract called for DOC to fund construction of Landsats 6 and 7. DOC would also reimburse EOSAT for operating costs of the old satellites over the course of the next 3 or 4 years, after which Landsat was expected to operate as a purely commercial business. Through good management, however, both satellites have operated nicely for many years. The contract also called for EOSAT to take over marketing the raw data from the Landsat series; it also specified that EOSAT would pick up the operating cost of Landsat 6 when it goes into service late in 1991. EOSAT would also pay operating costs for Landsat 7.

In spite of the contract, OMB has apparently not accepted the fact that EOSAT was to receive government funding for several years before being asked to stand on its own feet. Each year since the program began, OMB has balked at going to Congress for the money to cover what NOAA is supposed to pay EOSAT to keep Landsat alive. Looking at the words in the contract which it has with DOC, EOSAT rightly assumes that it will receive annual payments to cover the costs of operating Landsats 4 and 5, completing development of Landsat 6, and building the new ground station. EOSAT typically has waited several months for the final outcome of the annual federal budget cycle to know how much money it can spend. Such shenanigans partially explain Landsat's slow progress toward true commercial success.

It is interesting that, no matter which political party is in power, Landsat remains in trouble. There always seems to be a cadre of people in OMB who believe that it is bad enough for the government to have to pay to support weather satellites, let alone having to pay the cost of Landsat. The commercialization of Landsat was supposed to settle the question once and for all, and yet the question remains. Perhaps when the old satellites die and the new one has taken its place, we can put the period of bickering about Landsat costs behind us.

A good example of what EOSAT has been up against took place in early Spring 1989. Because the OMB had not requested money for NOAA to spend for Landsat operation, NOAA told EOSAT that no money would be available after March 15. Did NOAA care whether EOSAT went out of business? Based on what it heard from NOAA, EOSAT was preparing to close its doors when a group headed by the chairman of the newly formed National Space Council rounded up enough support to keep Landsat going for a few months until a more permanent plan could be generated. Some people saw this episode as a ploy by NOAA to avoid having to pay for EOSAT operations. They said NOAA is willing to go to the brink to see who the real supporters of Landsat are. There may be some truth to this. Unfortunately, the combination of NOAA's tight budgets and OMB's opposition to supporting Landsat have greatly complicated the transition of the program from government to private management.

A similar situation arose in 1990 when construction funds for Landsat 7 were necessary; again there were delays in finding the money. OMB's actions clearly reflected its frustration that the program is not yet self-sufficient. Why should the government have to subsidize a company that has a monopoly on the market? Why doesn't EOSAT go out and sell enough data to carry all the costs of the program? Unfortunately, raw data sales have never been large enough to carry all the costs of operating Landsat. The government certainly does not want EOSAT to solve its own problems by trying to take over the whole market for value-added products—thus putting out of business all value-added companies, many of which are really mom-and-pop shops. Adding to EOSAT's marketing troubles is the fact that the actions of OMB and NOAA in the past have made Landsat customers very nervous. Will the system be kept in operation or allowed to fall over the cliff during one of the frequent cliff-hanging exercises EOSAT has been subjected to? This uncertainty only encourages customers to switch loyalties to SPOT. Although some people may fault EOSAT for not being a more aggressive marketer, it is no wonder that EOSAT has had trouble building a successful business under the conditions in effect.

While EOSAT has presumably borne some of the costs of Landsat market development itself, and is also investing in building a new ground station, OMB seems to believe that EOSAT ought to do more to bring the program to the breakeven point sooner. In retrospect, I believe DOC failed to drive the right kind of bargain with EOSAT; it should have had a clear understanding of how much of the contractor's own money would be invested in the business. Recently, Peter Norris, Executive Vice President of EOSAT, told me that if recent growth patterns continue, Landsat could reach breakeven before many more years. It is not clear just how strongly EOSAT is supporting the program with its own money to try to reach breakeven sooner.

EOSAT enjoys annual Landsat data sales of about $25 million, with sales growing at about 20 percent per year. Norris thinks that when Landsat 6 goes into operation in 1991, sales will soon double. Can Landsat be operated profitably at that point? I think it will be at best marginally profitable, depending on the amount of ingenuity EOSAT can marshall to carry out the program. Data sales will certainly grow—perhaps by an order of magnitude, if we wait long enough. The real question is what will happen in the next 3 or 4 years while DOC is helping carry the costs.

Of course, the answer on whether Landsat can break even hinges on just how much revenue is needed to support such a program, including replenishing the satellites as necessary. The cost includes the operating cost of about \$20–25 million per year. Landsat satellites cost about \$150–175 million, including launch. If the typical satellite lives 5 years, we have an annual cost for operations and new satellite procurement which totals about \$50–60 million per year; this must be offset by sales of data. If it is possible to get this cost down to, say, \$40 million per year, the system stands a much better chance of operating profitably. If we can believe EOSAT's estimates of revenues of \$40 million in about 1994, and a continued upward slope of 20 percent per year for 5 years, a prudent businessman might still not be willing to invest in such a proposition, but it would at least make sense to see what further cost reductions could take place. To operate at a profit, one would need to reduce operating costs to less than \$20 million per year and allocate no more than \$20 million per year for satellite replacement. There is thus a great premium to be had by designing a less expensive satellite which will live longer—a task which EOSAT's parent companies are well equipped to do, and it would help to find a cheaper way to launch it. All this may be possible, but we do not know yet whether it can be done.

An alternative that appeals to me is to operate a satellite that serves the needs of two different kinds of users, thereby reducing costs by prorating costs of launching and operating between those two users. Later I will discuss who these customers might be.

Increasing sales would obviously also help reach the objective. Increasing sales will depend on how many new uses for the data are developed. An example of such a use is Mediasat, a hypothetical system of the future brought to mind by the sale of pictures taken by Landsat and SPOT when the Chernobyl nuclear power station in the Soviet Union suffered a catastrophic accident. The accident led some to envisage a growing market for satellite images. Mark Brender of ABC TV News went to the trouble of copyrighting the name Mediasat because he thought that such a system would be needed sooner or later by the news industry.

There may be such a market eventually, but many factors are working against it now. A strong deterrent is the fact that most newsworthy events happen in readily accessible places. If one can drive a car to the scene of the accident, or fly an airplane over the site, pictures taken with handheld cameras will be greatly superior to satellite photos taken from hundreds of miles away of an area that may be obscured by clouds at the critical moment. Another problem is that there just are not enough Chernobyl accidents to support an industry such as Mediasat. The Three Mile Island accident, for example, created no demand for satellite photography, since people on foot, in cars, boats, and airplanes could do the job. And even if we had used satellite photography at TMI, one or two accidents per year do not justify satellites which cost hundreds of millions of dollars to build and operate. It is highly unrealistic to assume that an occasional startling photograph can somehow cover the operating costs, much less the investment cost, of a Mediasat. I do not think Mediasat can be a viable system any time soon, or that its availability has much economic importance to the newspaper business.

But we should not discount the importance of seeing satellite photographs on the nightly news; this public exposure may lead to their being called on for something other than nuclear accidents, devastating floods, and similar dire catastrophes. If satellite pictures receive more publicity, perhaps this will help develop the market more rapidly.

Some have suggested that the U.S. government could save money by combining the needs of the military with the civil requirements and fly just one set of satellites. At the end of 1989 OMB proposed that Landsat should be transferred from Commerce to the Department of Defense—just the opposite of what one would expect in these days of *glasnost*. While this move has theoretical appeal as a way to save money, it may not be practical. The military always wants what the state of the art allows, even if it adds to the cost; making Landsat a commercial success means keeping it as simple as possible. Commercial use implies designing to cost and market requirements, leaving out the goldplated touches.

■ HOW TO SAVE LANDSAT

It would certainly help Landsat succeed as a commercial venture if we could come up with some large sources of revenue. We need totally new sources rather than slight expansions of existing customer bases. Two possibilities come to mind. For about eight years, several people, including myself, have been proposing putting the sensors that now fly on weather satellites on the Landsat spacecraft. That way, the Landsat operator would keep his existing customers but would also add the National Weather Service as a major customer. NWS would continue to get all the data now received from NOAA satellites but would buy the raw data from the Landsat operator. The Landsat operator would have enough income to stand on his own feet and would need no government subsidy. Right now, the U.S. government pays twice— once for collecting weather data and again for assisting EOSAT in covering its costs. Why not cut out one of those two costs and just pay the Landsat operator to collect the weather data?

The rather obscure objections I have heard to this plan seem to be based on the belief that it is some kind of plot to make U.S. citizens pay for weather data they now get free. The average citizen does not get information about the weather from those who operate satellites, and the NWS does not generate satellite data on its own. Rather, it processes satellite data collected by another agency (NESDIS), which taxpayers must support. The Landsat operator could take over certain appropriate parts of the work of that agency. This proposal certainly would not cost taxpayers more, and it very likely would cost them much less.

Another objection has to do with carrying several payloads on one satellite. If one fails, do you immediately launch another satellite or wait until more failures occur? Detailed analyses of these contingencies would have to be made, but I believe they would come out favoring the proposal to combine weather satellites and Landsats.

One might ask why the government did not see the economics of this approach at the beginning. The answer is that the two types of satellites arose at different times through different agencies, to serve different customers. No operating agency would benefit by a merger, even though the government would save money overall.

Another possible way of commercializing Landsat remote sensing is to collect the data under an international arrangement of some sort. Whereas the value-added phase of processing remote-sensing data is a cottage industry, lending itself to widespread entrepreneurship and a broadly dispersed workforce, the actual collection of data from space depends on flying sophisticated satellites that tend to get more complex and expensive in order to carry more and more sensors. This tendency obviously helps to drive up the cost of the satellites. I believe that these satellites are candidates for joint operation by several countries, if not for joint ownership.

One of the most attractive ways of internationalizing the Landsat business is to form a joint venture with other countries that need to collect essentially the same kinds of data as we do. This idea—or at least the idea of making agreements on what data to collect, what data formats to use, how to calibrate detectors so we can understand each other's data—is beginning to achieve a reasonable amount of support. A good prototype for how to create such an arrangement is given by the Inmarsat example. Drawing on the Inmarsat experience, perhaps countries already operating remote-sensing satellites could make arrangements to collect data for groups of nations (this was how Inmarsat got established; it leased capacity on other people's satellites). Of course Intelsat's charter includes collecting weather and remote-sensing data if two thirds of its governments agree.

Recently, an organization known as SAFISY has made progress on getting countries to plan for the sharing of data on some kind of joint basis. SAFISY—the Space Agency Forum for the International Space Year—is an informal but effective coordinating body of senior space agency officials of 24 countries which has met several times in the past two years to coordinate plans for collecting data for Mission to Planet Earth (see below). Ideally, those 24 countries will come to agree that the economics of data collection suggest an international cooperative approach to the problem. While much progress has been made on getting these countries to agree on technical data requirements, it may not be so easy to get agreement on all the details of the collection process itself and on possible ways of creating and managing an international data-collection system.

Reports circulated in 1989 that CNES, the French space agency, and NOAA had talked about combining their efforts on SPOT and Landsat in some way. In the beginning, the French wanted to fly SPOT on their own, both to develop the technology embodied in SPOT and to use the satellite as a prototype for a higher resolution military system in the planning stages. Now that their original objectives for flying SPOT have been accomplished, the French may be ready for a new agenda. In fact, I have been told that they approached the United States several years ago during the SPOT development phase to explore whether we would consider a joint venture of some sort, but found no interest here. Since it costs the French roughly the same amount of money to fly SPOT as it costs us to fly Landsat, it seems that each of us could save about half the money if we joined forces and flew only one set of satellites. Perhaps we could take turns buying new satellites to replace dying ones, or encourage two teams of contractors with French and American members to compete for the construction of the next SPOT/Landsat satellites.

SPOT II was launched in February 1990, and SPOTs III and IV are already well along in construction. If all these satellites are launched successfully, they will carry the French remote-sensing program until about the year 2000. After that, perhaps a joint venture will be feasible. If so, we should begin making plans now, because it takes 5 or 6 years just to plan and execute a program of satellite construction, independent of the complications caused by forming a new joint venture between nations.

But how does the concept of internationalizing Landsat comport with the concept of commercializing it? Fortunately, the Intelsat and Inmarsat consortia are models by which to answer. In both cases, member countries are free to designate an entity of their choice as their representative in the consortium. Most countries chose to designate the appropriate national PTT telecommunications organization, but the U.S. government designated Comsat, a private company subject to government regulation. There is every reason to suppose that if a remote-sensing consortium were formed, the United States would designate another private company as its representative. Intelsat and Inmarsat were set up to operate primarily in the communications business, and it is not obvious that remote sensing can ever rival that business in income and earnings. But it is worth noting that several Intelsat 5 satellites carried a maritime package for Inmarsat, with Inmarsat paying only the incremental costs. This was a real break for Inmarsat. Why couldn't these same satellites carry weather packages also?

One hundred countries need more or less the same data now collected by Landsat and SPOT. At least 24 of them are planning to put up Landsat-like satellites to collect their own data, even though many of the 24 cannot really afford to spend the large amounts of money involved in the construction and operation of their own systems. A consortium of 24 countries could achieve great economies by banding together; the savings would be even more dramatic if a consortium of 100 countries were to take on the task.

But saving money is not the only issue involved here; in addition to policy issues, there are prestige issues as well. So rather than try at first to get all nations to agree, we might start with just two. The U.S. and France understand each other, have cooperated (and fought) in their roles as members of Intelsat and Inmarsat, and have years of experience in the operation of two fine remote-sensing systems; they may be the ideal partners to start the ball rolling.

While some customers want Landsat data as such, others are willing to go elsewhere if they can get comparable products covering the area of interest at less cost and at comparable or better quality. Many of them want to continue with Landsat to maintain real or imagined compatibility with older files; others may drift to SPOT. (There is a slight difference in data format between SPOT and Landsat, but it was minimized by coordination between the French and the Americans. Processing systems are now available that work with both. Data products are available from both at essentially the same price.)

■ CONDITIONS FOR SUCCESSFUL COMMERCIALIZATION

Can land remote-sensing systems be operated successfully on a commercial basis? To answer that question, first we must determine if there is a pent-up demand for Landsat-like products. The answer is yes. People who have been using Landsat data for mapping, land-use studies, and so on will need such data for the indefinite future. They have been able to get Landsat data for 15 years at very low prices and want to continue to do so. But we must distinguish between Landsat and Landsat-like data. Landsat satellites were designed more than 10 years ago; they were not designed with commercial customers in mind and their technology is not the latest; we would do better to start with new designs optimized for commercial as well as government needs.

Second, is there a stable market for Landsat products? The federal government is and will likely remain the largest customer for Landsat data. There is a perishable quality to much such data—for example, that used for crop analysis, area flooding, snowfall buildup, and urban growth—so the same customer must buy frequently as his files become obsolete. The market for this kind of thing should be reasonably stable—especially since various government agencies have used Landsat products for many years and should know what their real needs are. Aggregating federal needs could help.

State and local governments are the second and third largest users

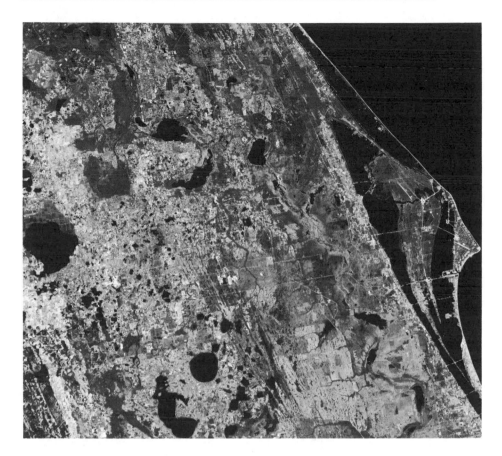

Figure 14. Orlando, Florida, and nearby Cape Canaveral, as viewed from Landsat. (Photograph courtesy of NOAA.)

of the data. Neither they nor the federal government have made projections of their needs. If EOSAT knew what the level of purchases for data would be from its various government customers, it could do a much better job of planning and have a rather good idea of the average demand for its products. Such data would let it project whether Landsat 6 can be a viable commercial venture. As things stand now, EOSAT has to wait for months at the beginning of each year to find out whether it is supposed to close the door on the whole operation or prepare for a buildup of business. We have had one bit of good news lately, however. At a recent meeting of the National Space Council, the chairman, Vice President Quayle, was able to get agreement reaffirming a long-term U.S. commitment to support the Landsat program.

Third, can space be competitive with other sources of data? Certainly aerial platforms are more cost-effective than space platforms for remote sensing in certain cases—if the area of coverage desired is small or localized. But only information from satellites allows us to construct a global picture of the state of the wheat crop worldwide or the extent of blight of a certain kind. Sensing via satellite is less intrusive than sensing via aircraft; sometimes nations—who retain the right to deny access to airplanes which they think may be snooping—may not permit aerial photography. The Chernobyl accident was a case in point where airplanes were not permitted but satellites could not be denied access.

Fourth, what is the state of competition among different suppliers of the same products? The present competitive picture is unclear. SPOT Image has no profit-and-loss responsibility for SPOT, since SPOT costs are covered in the CNES budget. SPOT Image's job is to sell as much data as possible; it will get a gold star if it can cover the cost of operating the so-called ground segment. It might be judged successful no matter what revenue it generates, since any revenue represents an offset of SPOT's cost of operations. (SPOT Image must pay a small royalty to CNES for the data it sells, but the amount is trivial.) EOSAT, on the other hand, has to begin earning money by the time the government delivers Landsat 6 and a new ground station; at that point, the government will have fulfilled its contract.

In 1990 Congressman James Scheuer introduced a new bill in the House which calls for using the Landsat system to conduct a world-wide survey of forests. This would be a permanent job for Landsat—that is, a continuing survey is called for as forests are cut down and replenished, with the objective of the whole exercise being to stimulate people everywhere to become forest-conscious, being careful to replant more than they use. If Landsat is that good—and if the bill actually passes, which is far from certain—it only takes a few such jobs to ensure Landsat's future. Is this commercial business? Probably not, but it will do until more commercial business comes along.

As for international competition, it is heating up, with a number of countries getting ready to launch satellites comparable to Landsat.

This makes the odds rather good that Landsat, SPOT, and Soviet data sources will be joined by several others fairly soon, including the Japanese, Chinese, Brazilians, Indians, and satellites from the European Space Agency. In a field where data sales have been about $40 million per year, there has not been enough revenue to keep both Landsat and SPOT healthy. Even if sales continue to grow rapidly, it will be hard for anyone to earn money after a dozen or so suppliers appear on the scene. Of course, not many of the 20 or so newcomers are planning to go into business to make money. Many are doing it for scientific reasons. They want to be able to collect data about their resources and their environment, and some of them do not want to be dependent on foreign suppliers. Several countries are probably doing it for reasons of industrial policy; they want their own aerospace companies to produce space hardware and software. Several may be doing it for the same reasons the French are doing it; any money they collect from data sales will offset some of the cost of operations, no matter why they built their systems in the first place.

The old specter of competing with governments is raising its head again in this case. Can a private company—EOSAT or anyone else—succeed in commercializing Landsat if SPOT operates under different rules in competition with it? While we are trying to commercialize Landsat as a system, including constructing and flying the satellites, the French merely want enough customers to cover the cost of the ground system. So our government's plan—commercialization in competition with several governments—may not be achievable. Perhaps by the time plans for constructing replacement satellites for the present SPOT satellites are being drawn up—roughly 1995—the French and U.S. governments will have agreed that a joint effort of some sort is good for all parties. Failing that, the only hope I see for successful commercial operation of Landsat is for the satellite to start collecting weather data.

Various news media—including the September 1990 issue of *Via Satellite*—carried the story of John Knauss, head of NOAA, testifying before Congress about the lack of success of the Landsat commercialization initiative. He believes no commercial success will be achieved in the next one or two decades. His testimony supports my statement about Landsat's inability to make money under present circumstances. Nonetheless, there have been some encouraging signs lately which make me hope Landsat may eventually become a successful commercial venture. NASA is reportedly considering buying scientific environmental data from private companies. Perhaps private companies could collect certain data for NASA by putting special satellites in orbit or by mounting appropriate sensors on satellites they were planning to fly for other purposes. A Landsat satellite of the future may be able to enter this market also, adding NASA as a customer for certain special data.

Commerce Business Daily for June 20, 1989, contained an announcement from the Department of Commerce asking industry to develop options for ways to continue Landsat operations beyond Landsat 6. Among the alternatives the announcement suggests for continuing operations while minimizing expenses are government-industry partnerships (flying privately owned sensors on government-owned platforms) and multinational partnerships of U.S. companies and foreign governments. DOC stressed that these are only examples of alternatives which might be worth exploring. The implication is that Commerce is committed to continuing Landsat in some form but is searching desperately for a way to do so that will be both interesting to industry and more economical than anything that has yet been tried. I find this cause for considerable optimism.

■ PEACEKEEPING SYSTEMS

Space News for April 2, 1990, reported that members of the Western European Union have been examining for about two years the merits of building what are called treaty verification satellites. This idea is similar to a French proposal at the United Nations in 1978 to create an International Satellite Monitoring Agency. ISMA, in turn, is similar to an initiative suggested by Howard Kurtz, a Washington peace activist who uses the unlikely name War Control Planners. Kurtz's idea was to create a worldwide agency of peacekeeping by monitoring and publicizing all military moves by all nations. The theory behind all these proposals is that shedding the cold light of day on aggressive actions will build up public outrage and cause offending nations to curb their improper impulses. Whether this is true is not obvious; there have been many reports in the press about the use of satellites for monitoring the military situation in the Middle East since August 1990. The fact that the satellites were in use did not stop Iraq from invading its neighbor Kuwait. Of course the actual photos from these satellites have not been released to the public; perhaps they could have had a deterrent effect if they had been.

The WEU members were advised by industry to get on with building the satellites—a normal industry reaction whenever interesting and profitable work may be available. The governments were reported to be moving more slowly, suggesting instead the formation of a European agency to procure existing commercial data from SPOT to be used to verify arms control agreements now in existence. The French suggested that they could train photo interpreters to do this kind of work, drawing on their background in analyzing SPOT products. A French delegate used SPOT images of French military installations to show that, although it was desirable to have higher resolution data, adequate information was already available (through SPOT) to do a reasonable verification job.

While there was consensus in WEU to take seriously the formation of the agency to analyze appropriate data, the suggestion to form an agency to buy and operate satellites was turned down. Rather, it was decided to do a study of requirements and capabilities and to issue a report at yearend 1990. This plan has great appeal and in time such schemes may well grow to form the basis for cooperative satellite procurement and operation.

The difficulty of raising the quality of data to be shared among many nations for peacekeeping purposes was illustrated by an exchange with a French delegate. A proposal to obtain data from Helios, a military reconnaissance satellite under development by France, Italy, and Spain, was vetoed with the statement that the data will be classified and treated as such. With such sensitive information at stake, worldwide or even regionwide plans can be started only slowly, among a few nations that have a history of cooperation and alliance; the participating group can be gradually expanded as confidence grows in what will be done with the products. Helios may well turn out to be such a system.

▪ MISSION TO PLANET EARTH

Only a few years ago, most people did not give much thought to the environment beyond worrying about the three-day weather forecast. Recent public opinion polls show that three fourths of the people in the United States now think that we are in a serious environmental crisis. There is a surprising degree of consensus that the problem is real, that we must change not only our ways of living but also our ways of thinking about our world. Most people also feel that we need both new policies and new leadership to deal adequately with these important environmental issues.

Our space assets are a vital link in the process of studying our environment. Space is the place to gather a major fraction of the data we need to understand the earth for the first time in all its beauty, complexity, and increasingly recognized vulnerability. In 1988 the National Research Council's Space Science Board carried out a series of studies under NASA sponsorship with the general title "Space Science in the Twenty-First Century." A subgroup, the Task Group on Earth Sciences, called their report "Mission to Planet Earth," to highlight the need to focus at least as much attention in our space program on understanding our home planet as we do on understanding other worlds—even though exploring other planets may seem more exotic and intriguing.

Mission to Planet Earth as used today is the name of an overall plan to use space-based technology to learn how the earth works as a system. It includes the study of "natural" changes that are due to internal processes in the earth, the effects of cosmic radiation, and the influence of living organisms, as well as the "artificial" changes caused by humankind. The mission would establish a global observation system in space that would take long-term measurements of a wide range of earth processes. A comprehensive data system would be organized to allow quick and efficient access to large volumes of data at various levels of processing, and elaborate computer programs would be developed to allow modeling of various systems and eventually the prediction of future changes in the environment.

For millions of years after they evolved, humans did not have very much effect on the world. About 5,000 years ago the grazing of domesticated animals and the cultivation of crops began to cause erosion and other stresses on the natural arrangement of things, but it was not until the Industrial Revolution just 200 years ago that humankind began to have a significant impact on global systems.

How can we sort out the environmental changes for which we human beings are responsible from those awesome changes caused by tornadoes, hurricanes, earthquakes, tidal waves, volcanic eruptions, and plagues of locusts—all of which occur without our intervention? Evidence from tree rings, ice cores, and lake bed sediments, for example, show that the earth has been cycling through a wide range of climates and other upheavals over the eons of the past. Plate tectonics and movements of the earth's crust, interactions of such movements with volcanoes, ocean waters, and minerals, changes in the earth's magnetic field and its interactions with the solar wind, natural variations in trace gases in the atmosphere, variations in the earth's orbit, its tilt, and eccentricity, and changes in the solar constant—all these have been partly responsible for major climatic changes in the past, including the recurring ice ages. Undoubtedly other natural factors affect climate besides the ones we think we know about. We cannot hope to deal with the issue of human effects on the environment without first having a much greater knowledge of the many interactions that take place between living things on earth and all the natural phenomena that proceed on their own.

Before we are ready to do anything other than being generally careful about energy conservation, reducing birth rates, stopping topsoil erosion, and a few other things, we have to first understand the earth as an "inanimate" object, then learn about the effects of living things on it, and, finally, confront the effects of people as our numbers and perceived needs expand. Laying out an adequate program to study such changes in the physical world and their interaction with the world of living organisms, especially humans, is a major task.

Because spacecraft can collect global data more quickly than any other means, remote sensing could well be one of the 3 or 4 principal foci of NASA's activities for the 1990s. Carrying out Mission to Planet Earth is a way for NASA not only to do necessary work which no other U.S. agency can do but also to confirm that NASA is interested in what might be called real-world problems. Focusing on Mission to Planet Earth would strengthen considerably the support of NASA by those who now applaud the agency's role in increasing our understanding of the universe but do not see NASA doing much to improve our everyday lives.

■ THE INTERNATIONAL SPACE YEAR

In 1985 Senator Spark Matsunaga proposed that the year 1992, the 500th anniversary of Columbus' voyage to America, be designated the International Space Year. His vision was that the ISY would be a time of enhanced use of space by many cooperating nations to achieve a better understanding of our world and our universe. His resolution was endorsed by both houses of Congress and signed by President Reagan. Matsunaga also caused the creation of US-ISY, the U.S. Association for the International Space Year. US-ISY in turn held conferences jointly sponsored by NASA to explore various space initiatives on a multinational basis. At one such conference, in April 1988, the Space Agency Forum for the International Space Year (SAFISY) was formed; the space agencies of 17 nations participated in the initiating conference. The group has now grown to 24 nations. SAFISY is a mechanism for convening the heads of space agencies and their scientific and technical representatives to coordinate activities on Mission to Planet Earth.

Impressed by the immensity of the task of environmental data collection and analysis, the conference proposed setting up pilot projects in a few selected fields. This would enable cooperating countries to work out detailed mechanisms to achieve successful results before tackling the full range of environmental issues. For example, the Greenhouse Effect Detection Project would collect detailed data on local and regional temperature averages and extremes, making comparisons of what is observed with what the best models predict.

The conference emphasized the fact that data collection is not an end in itself; rather, the objective is to derive information about our planet that will enable us to make decisions. All this collection effort must be accompanied by development of more realistic computer models and extensive interdisciplinary analysis, with results compiled in various ways to provide proper understanding of hundreds of essential parameters about the environment as seen from a global perspective. Thus data collection is only the first step in a chain of related events; data processing and analysis are perhaps the most difficult and time-consuming tasks in the chain.

In the space field we have regularly had problems handling the massive data generated by spacecraft; this problem will be exacerbated by the arrival in 1998 of the Earth Observing System (EOS), whose two sets of polar-orbiting platforms will hold dozens of sensors capable of collecting high-resolution data at gigabit rates. Space enthusiasts sometimes forget that collecting data is not the end of the exercise but only the beginning.

The SAFISY conference established several working groups, one of which compiled a list of all the spacecraft various countries were planning to launch over the next decade to study the earth—land, water, and atmosphere. Most of us were surprised to see 48 satellites on the list. This was the first time the major participants had ever seen the total list of remote-sensing satellites pulled together. About half the satellites on the list were designed to create images of the earth—that is, to make what look like pictures even though they may not be made with visible light. Such satellite sensors are good for studying and predicting weather and for looking for currents and life forms in the ocean, pollution in streams and lakes, blight in crops and forests, snow buildup in mountains, the aurora over the poles, and many other interesting and important phenomena. The other half were designed to measure radiation in such a way as to determine the amount of water vapor and various other gases in the atmosphere and the temperature of these gases at different altitudes.

It is gratifying that the momentum established at the first SAFISY conference has been maintained. A second SAFISY conference in May 1989 in Rome and a similar conference in May 1990 in Tokyo demonstrated no flagging of zeal on the part of participants. On the contrary, many, including the Soviets, seem more dedicated now than ever before to dealing seriously—and on a global basis—with the problems that we all must tackle together.

■ THE ROLE OF DEVELOPING NATIONS IN EARTH OBSERVATION

Because satellites, being several hundred miles above the earth, are somewhat crude in measuring vertical profiles of various gases and other phenomena, each country must be involved in studies to fill in the necessary details. Then by integrating such data with space-derived data, we can create the complete mosaic of what we need to know.

All of the space data must be calibrated by making direct in situ measurements within the atmosphere, on the ground, and in the oceans. Space-based technology cannot teach us everything we need to know about the earth. Some important parameters are not easily measured from space, including precipitation, ice thickness, heat fluxes over land and sea, and cloud radiation. These must be determined through measurements made at the specific locations of interest by instruments that have been accurately calibrated and installed by people on the site.

While a few spacefaring countries can by their own efforts collect the most necessary space-derived data, we must rely on every country to supply us with data collected locally—on land and in airplanes, sounding rockets, balloons, shore stations, ships, and buoys. Such measurements are essential if we are to determine accurately what is called "ground truth"—the true conditions that prevail on the surface, in the air, and in the oceans. Every country can afford to collect ground truth. To reduce the cost of collection of the data, and to feed back to all countries the bigger picture that will help them understand climate and other effects, what we need is a partnership among those few countries operating in space and those many which will not go into space but which must be a part of the worldwide data-collection effort.

For a number of years, the United States has operated a system whereby ground-data collection platforms transmit the data they collect to stationary satellites that in turn relay the data to a central collection point. Satellites that perform this relay function are the same GOES weather satellites that collect the information we see nightly on TV news. Such a system could conceivably be developed on an international scale, in the form of a partnership between countries operating such satellites and all other countries.

For example, a Third World country might install a data terminal in a major river to sense and record several interesting parameters such as water depth above mean level, salinity, acidity, bacteria count, temperature, flow rate, water quality, and turbidity. The data terminal would incorporate a receiver-transmitter (a transponder). When a suitable satellite sends a query to the ground station (data terminal), the ground station transmits the data stored in its memory over the last few minutes or hours to the satellite. The satellite in turn relays the data to a central data-analysis station in the same country, if it exists, or to a mutually acceptable data-analysis center in another country. The data may enter the public domain, and thus become available for all people who want to monitor such processes on a worldwide basis. Satellites capable of operating in this mode are now in position all around the world and can collect the data from ground stations and other instruments in and around all countries. They thus could have an important role in maximizing global data collection while minimizing cost.

Another powerful illustration of international cooperation in earth observation involves weather balloons and was practiced on a limited basis a decade ago during the Global Atmospheric Research Program (GARP). A transportable box is prepared in a space-capable country and then shipped to a country which agrees to cooperate in collecting meteorological data of mutual interest. The box includes an array of atmospheric sensors, a transmitter, a balloon, and a bottle of helium. At the appropriate time, someone (not necessarily trained in meteorology) opens the box, causing the balloon to inflate and ascend into the atmosphere. The sensor package measures the desired parameters and transmits signals to a satellite, thus giving all the sensed information about the atmosphere as the balloon ascends to higher altitudes. This is an excellent example of how countries of vastly different technological capabilities can join in an activity that gives to each something very important: the country using the balloons gets the information it needs about the local situation; the world meteorological community gets the data to round out its picture of worldwide weather phenomena.

A refinement of this includes a Global Positioning System receiver. The transmitter that broadcasts the signals from the weather sensors also broadcasts the GPS signals received on the balloon. This allows the central data-collection point to track the balloon's position and velocity, giving an accurate measurement of winds aloft as the balloon ascends.

Another example applies to the UHF Doppler wind profilers. These relatively simple low-powered short-range radars pick up back-scattered signals from small particles blowing in the wind. They read out wind velocity as a function of altitude. The results may be picked up by a GOES or similar satellite for relay to a central analysis point.

Measurements by instruments detecting trace constituents in the atmosphere could be operated in similar fashion. The problem is to make the instruments reliable enough to be operated routinely by relatively untrained people.

Another instrument that could be installed inexpensively in many countries is an automatic raingauge. This is an optical device which measures attenuation caused by rain. This signal can be sent to the satellite from thousands of installed raingauges, yielding much valuable data not now received about rainfall and not readily measured by satellite. Technicians in the host country should be able to maintain the simple instruments contained in such devices. Soil moisture can be measured by various instruments including temperature measuring devices, the results of which can be relayed to the GOES weather satellites.

Thus there are many ways that less developed countries can play an extremely important part in data-collection efforts. Space assets operated by a few countries plus ground truth supplied by many others can give us the worldwide picture we need. In some cases, the equipment to collect the data may be supplied at no cost to the host countries by the spacefaring nations which need the resulting data to round out their own information on worldwide environmental concerns. In other cases the equipment may be purchased on the open market.

This leads to the question: What role can the private sector play in implementing the many projects of Mission to Planet Earth?

■ THE PRIVATE SECTOR IN MISSION TO PLANET EARTH

Many factors point toward an enlarged role for the private sector in Mission to Planet Earth; other factors are not so encouraging. Favoring private initiative is the state of the federal budget and deficit, which leads some people in the federal establishment to see private investment as a way to avoid large public expenditures while still responding to the environmental crisis. This attitude was evident during the Reagan years when people proposed that certain major components of the Space Station could be leased rather than bought. Most of us viewed these tendencies with a jaundiced eye; did it make sense to lease the remote manipulator arm fulltime for 30 years rather than buy it? Undoubtedly, the answer was no. But there were other suggestions—having the private sector build and operate the neutral buoyancy facility or swimming pool, for example. This facility could be readily available at a NASA base where one can expect dozens of private companies to find uses for it. Having NASA lease time in it for experiments, while a major fraction of the cost is amortized by private users, makes sense—at least in theory.

The same principle could apply to private companies' collecting satellite data on NASA's behalf. NASA expects to accomplish a major fraction of Mission to Planet Earth using the very expensive Earth Observing System satellites. EOS satellites benefit greatly by having a dozen sensors that can take many measurements from the same orbit, but they are penalized by high unit costs. Because so many eggs are in one basket, when a satellite is lost, it is a major loss. Clearly the private sector cannot afford to underwrite the cost of an EOS satellite. But there will always be a sensor that should have gone on the latest EOS but was not quite ready when the rocket was launched. That sensor can be flown on a private satellite that happens to be going up at the right time and has the power capacity and room to carry the extra sensor. All that is required to get the private sector involved in this way is an open attitude on the part of NASA that it will not arbitrarily prevent such actions when they make sense. In appropriate cases, a data stream can be collected and processed by the private sector, then fed to NASA in the proper format.

About 20 ground stations now receive data from Landsat. Several of them have been modified to receive SPOT data also. It is a logical step for those same countries to generalize their data-collection efforts and to modify their equipment to receive data from the many remote-sensing satellites that are expected to operate in the future to collect information about the earth. If China puts such a satellite in orbit, why should not those 20 ground stations already installed in various countries be modified to permit them to receive data from the Chinese satellite? Certainly, it becomes easier to receive a third data set after one has already mastered the art of building a ground station that receives data from two sources. These terminals have extended the fraction of the earth's surface from which data can be received, but the terminals cost several million dollars each. The private sector could build such terminals and operate them, supplying a data stream at a certain monthly cost.

Not all of the 20 or so countries that have stated a desire to fly remote-sensing satellites have the technological expertise to accomplish this goal. The private sector may be able to accomplish what poor but proud governments have so far lacked the will to do, which is to operate remote-sensing satellites jointly for two or more customers. One can imagine a satellite that crosses over 4 or 5 countries at different times, sending to the ground a stream of data whose header changes to reflect the proper nationality as the satellite crosses the boundaries.

At the present rate of satellite launching, we will soon see the proliferation of land-observing systems operated by many countries, with many of these satellites being almost identical in their design and operation and thus quite wasteful of resources. Some duplication can be justified to assure receiving the necessary data in case of equipment failure, but having a dozen essentially identical systems in orbit cannot be justified by any logic. I have proposed that those countries planning to fly the same kinds of satellites should get together and agree on how to design, build, and operate satellites jointly, thus saving money.

Some of that money could be better spent analyzing data already gathered, or in setting up data centers where data of all types would be available. An aspect of Mission to Planet Earth that is receiving increased attention is the past neglect of data analysis; some people are even saying that EOS may be in jeopardy because of failure to give adequate attention to this problem. Since it is apparently more fun for engineers to collect data with exotic satellites than to process the data and try to fathom what it all means, there is a large and growing private-sector market in data analysis. Many companies have shown that the private sector can make a fortune in this field. It is not unreasonable to believe that the data-analysis industry can make an invaluable contribution to Mission to Planet Earth and probably become very profitable in the process.

The money saved by forming an international partnership in a remote-sensing consortium could also be spent to set up training programs at such data centers where people from Third World countries would be trained in solving problems in their own countries. Many countries that cannot afford to fly satellites could afford to apply the satellite-derived data to their own needs, if they had trained analysts. Therefore, a role for the private sector in the United States could be to train people for such work. I believe that it could be done better at centers having all kinds of data available—a kind of open market for environmental data where many countries would supply data they had collected, and people from many other countries would come to take advantage of the resulting "library." Perhaps the price for membership in the library should be related to how much data the member has contributed to the library and how much it has taken out.

There could be two types of global data consortia. One country might decide to join a consortium that was building satellites; the country would pay money to become a member and would get money back for equipment delivered to the consortium. The ESA countries do something like that today. That same country could also decide to join a library consortium, paying in money for data taken out of the library and being paid for data supplied to the library. Another country might decide to join only the second consortium if it had no plans to participate in building satellites.

Factors not auguring well for the private sector include the desire of governments—including our own—to buy all the hardware rather than lease it. If egos are satisfied only when everything is owned outright, there may be little room for the private sector other than constructing the hardware in the first place—a nonnegligible task. But if governments will accept that they can get the data they need without insisting on ownership, the private sector may not only build the hardware but also sell services to various government customers, with the possibility that no one government agency would have to pay the full cost of the operation. That way, all could come out ahead.

6

NAVIGATION

One of the oldest and most obvious ways to navigate is to use the heavenly bodies, most of which perform a rhythmic motion that is easy to describe and, best of all, is predictable. While the sun and moon can give valuable clues for navigation, we cannot use them alone to tell us where we are. The stars are different; by using combinations of stars, we can find our position and plot our course over the surface of the globe. But stars are useless in daylight and on cloudy nights, so long ago even primitive peoples realized they had to develop other ways to plot their course.

During my time as a part-time navigator on a Navy ship during World War II, I used standard navigation techniques that had served sailors well for hundreds of years: the sextant and chronometer. Weather permitting, we "shot the sun," using the sextant to read the height of the sun above the horizon at noon and the chronometer to tell us just when the sun rose and set, giving us both longitude and latitude readings. We also used dead reckoning—knowing from whence we started and the direction and speed since leaving there, we calculated our new position to within a useful degree of accuracy. As time went on, we developed ways to improve this very old technique. But it still depended on devices that aged, grew barnacles, or took little account of ocean currents, so we frequently found errors in navigation that added up to 30 or 40 miles. For that reason, we tried to update our position frequently by other means, using dead reckoning only to estimate how far we had come since the last "fix."

Radio devices served us much better. When we were fairly near shore, we would take bearings on several known radio stations and plot these lines of position on a map. We knew we were at the point where the lines of position crossed each other. Depending on various factors, we could find our location to about a mile accuracy. Another radio system we grew to like was LORAN (Long Range Navigation). This system depended on chains of radio stations, consisting of one master station and several slave stations that merely retransmit the signal they receive from the master station. By listening to all the stations at once with a special receiver that measures the differences in times of receipt of signals, we could find our position to about a mile accuracy.

A new and popular device at that time was radar, which would give position very accurately if there happened to be a readily identifiable landmark nearby; on the open ocean, it was useless. Radar beacons at known locations, and triggered by one's own radar, were also helpful, especially for airplane use. These beacons are like the navigation system called Distance Measuring Equipment (DME) now in use in air navigation. They operate on the same principle as an airplane's transponder, which tells the FAA air traffic control system where the airplane is.

Several new methods of navigation became available shortly after the war. The simplest was Omega, a variant of LORAN that extended use of the technique to the whole world. Its best accuracy is about one mile, with larger errors in certain locations. The early LORAN system later came to be called LORAN A and has now been replaced by the more accurate LORAN C. Another popular device was the inertial platform, combining a number of gyros and accelerometers to measure speed and direction. When first employed, these devices allowed errors to build up at a rate of a few miles per hour; more recent versions have reduced the errors significantly. These systems were heavy and expensive, and although their weight has been reduced, they still are not priced within range of the typical boat enthusiast.

Ships of course can afford to spend many thousands of dollars on precision navigational devices like multiple inertial platforms. Groups of three are typical; if two agree, their readings are accepted and the third is ignored. It is rather common to combine two navigational methods into one overall system which draws on the strengths of both types of sensors. For example, one may combine an inertial navigator, which has good short-term stability, with Omega, LORAN, or radar—systems that have good position-fixing capability but also have their own weaknesses.

■ TRANSIT

The best navigational devices invented so far are artificial satellite systems that cover the entire globe (with the exception, in the case of geosynchronous satellites, of a region near the poles). The Navy pioneered in developing space-based systems for navigation shortly after we mastered the art of putting up satellites. In the 1960s a number of satellites called Transit were placed in polar orbits about 700 miles high, where they became in effect artificial stars to navigate by. Transit transmits a very stable frequency and also time marks. The ship's receiver measures the frequency very accurately, and the operator can determine his position with respect to the satellite using the principle of Doppler shift. Keeping track of the rate of change of the received frequency tells the operator how close the satellite came to the ship at the point of closest approach. Since the track in the sky traced out by the satellite is known and is transmitted by the satellite to the ship's receiver, the ship's position is readily determined by processing the received signal. From a stationary position, Transit gives measurements to about a fifth of a mile accuracy. The accuracy degrades as the ship's rate of movement increases.

While Transit was a very impressive and useful system for its day, a major drawback is that the satellites are at low altitude, and so they spend most of their time below the horizon where they cannot be seen. A satellite in polar orbit, like Transit, will pass over a given ground position about every 12 hours. Because there are several satellites in different orbits, usually one will come by in one or two hours. Of course this could seem like an eternity if one happened to be in a storm and were in danger of being blown on the rocks.

The Transit system was developed by the Navy for the purpose of charting its ships' courses more accurately. The Navy maintained the ground control stations, put the satellites in orbit, and bought special-purpose receivers for Navy ships to use. But the signals transmitted were available at no cost to anyone with the proper receivers, and therefore thousands of civilians bought this equipment on the open market at a cost of about $1,000–2,000 each so that they could use the system for navigation. Pleasure boats especially found Transit a satisfactory way to navigate, since they can tolerate better than commercial boats the sometimes long delays between sightings.

The space segment of the Transit system, because it is operated by the Navy and available for anyone's free use, is not commercial in any sense. But the ground segment—which consists of the military and civil receivers—is commercial. The companies that make Transit receivers have sold thousands of them off the shelf to all kinds of customers. Other than commercial communications satellite systems, these sales were one of the first examples of sales of commercial products based on space assets.

Unfortunately for the owners of these receivers, the Navy's Transit system is being superseded by a more sophisticated one (developed and operated by the Air Force) which requires different—and more expensive—equipment. If the Navy abandons operation of its Transit satellites, as is expected, the owners of Transit receivers will be stuck with equipment that is no longer functional. But this will take a few years to happen.

■ NAVSTAR

The space system which is replacing Transit is called Navstar. It will cost billions of dollars to complete, but it has a tremendous advantage over Transit: the satellites are high in the sky (12,000 miles versus 700 miles for Transit) and their signals can be received by anyone on the open seas, on land, or in the air. (People in submarines need a special solution, however.) The space segment is operated by the Air Force, whereas the ground segment receivers are purchased by the various military services from commercial suppliers according to their individual needs. Navstar—also called the Global Positioning System, or GPS—seems to be the wave of the future in global navigation.

GPS is an extremely elegant system, capable of enabling a typical military user with the right kind of equipment to find his position anywhere on the earth's surface or in the sky to an accuracy of 10 meters. Under the right conditions, GPS can be used to measure distances to an accuracy of a few centimeters, by the use of what is called differential GPS measurements. If a GPS receiver is put at a very accurately known reference point, its position as measured by GPS can be compared with its known location. The differential between the two can be used to correct other measured positions in the general area. In addition to the obvious ways of using GPS for navigation, its extreme accuracy opens up the possibility of using it to measure such things as the subsidence of the land around Phoenix caused by pumping out ground water, or the rate of increase of the spread between the two sides of the San Andreas fault near Los Angeles. This capability may soon permit geophysicists to predict incipient earthquakes.

Such extreme accuracy also allows us to contemplate doing away with many—but not all—ground-based navigation systems, saving billions of dollars in the process. While we would not want to rely totally on any one system for all our navigational needs, we could certainly think seriously about replacing most aircraft location systems like VOR/DME (VHF OmniRange/Distance Measuring Equipment) or VORTAC and landing aids like ILS (Instrument Landing System), keeping perhaps the LORAN ground-based systems as backup for possible outages of Navstar. Recently, however, the Air Force has begun encrypting the more accurate Navstar signals, leaving only the secondary signals for civilian use; the Air Force has said those signals can always be counted on to give position within 70 meters' accuracy, but that would not be accurate enough to replace ILS. Consequently the ILS system is likely to be with us for some time, especially if the Air Force continues to insist on encrypting the high accuracy Navstar signals.

The Air Force became interested in Navstar as a result of the Vietnam experience. They had had difficulty installing a ground-based system in that area after hostilities started. Looking ahead to a time when they might need a good navigation system in some other faraway land, they wanted to put a system in place ahead of time, rather than having to install one on a moment's notice after some military or political conflict became imminent. In designing this global system, they made it "passive," meaning that airplanes could navigate accurately anywhere on the globe without having to radiate telltale signals that could be picked up by an adversary's receivers.

LORAN C and D, which had been set up in or near Vietnam, were passive and did the job of determining position rather well over a small area. But if an airplane got too far away from the transmitting radio stations, errors in indicated position grew to an intolerable degree. For that reason, LORAN transmitters had to be located near the battle

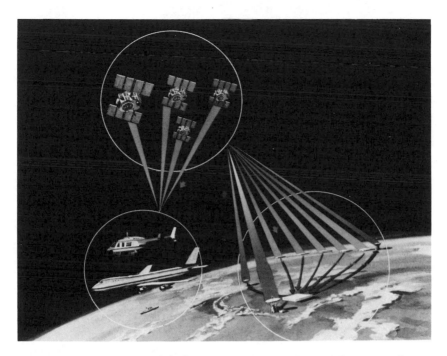

Figure 15. The Navstar Global Positioning System (GPS), with its constellation of 20 satellites, will allow properly equipped users to calculate their position to within tens of feet, their velocity to within inches per second, and the time to within a fraction of a millionth of a second. The key to Navstar's accuracy is a set of atomic clocks that lose or gain only one second in 300,000 years. Ground stations track the satellites to correct their positions and synchronize their clocks. (Photograph courtesy of Rockwell International.)

area. Unfortunately, transmitters near the battle area constitute a lucrative target for the enemy.

The answer to the problems of ground-based systems was a space system with the satellites located so high that very few, if any, countries could attack them, and with the control stations located in the heart of the United States and hence unlikely to be attacked. Navstar meets these criteria. With an eventual constellation of about 20 satellites (18 plus a few spares in orbit) at an altitude of about 12,000 miles, all controlled from stations in the United States, Navstar passes the vulnerability test. The satellites are deployed in 6 different orbits, each tilted 60 degrees from the equatorial plane. Signals from at least 4 satellites can be received at any location on earth at any given time, even though the satellites are all moving across the sky. As of September 1990, 8 of the 20 projected Navstar satellites were in place and offering service.

Navstar works something like an inverted version of DME. This air navigation system works by using radar-like pulses to measure the distance from fixed ground stations. The airplane's receiver computes its position by a kind of triangulation. If you are 70 miles from the Washington station, for example, 60 miles from the Richmond station, and 100 miles from the Norfolk station, there is only one place you can be. The first tests of the basic principle of the Navstar system and its inherent accuracy were actually carried out about 15 years ago by putting simulated "satellites" on the ground and flying the receivers over them. Those tests confirmed that Navstar had the required accuracy, not only on the ground but also in the air, since it operates equally well in three dimensions, giving altitude measurements of the same precision. Navstar can give accuracies of about 10–15 meters.

Each Navstar satellite contains four cesium or rubidium clocks, which are among the most accurate clocks in existence. The clocks are set when the satellites pass over the control station. By listening to the signals from these clocks, the receiver is able to calibrate its own clock. Then it measures the time that each satellite signal is received. From that information, it calculates its own position in three dimensions. Using its ability to measure frequency of the received signals very accurately, it calculates its velocity with high precision. The orbits of the satellites are measured by the control stations in the United States and are transmitted to the receiver with great accuracy also, so very little error exists in the system. To avoid timing errors caused by varying atmospheric refraction, the satellites transmit dual frequencies which are affected differently by the atmosphere. This difference allows corrections to be made, eliminating all but very minor errors.

Navstar is meant to be used by all the military services. The Army has developed back-pack and Jeep-mounted receivers, the Air Force has developed systems for use in the air and in space, and the Navy has its own designs. Different uses demand different accuracies and different parameters. A foot soldier and a sailor do not need to know how high they are; they are on the surface. Their receivers can be of simpler design than those that must compute altitude. By such techniques, a range of accuracies and prices can result. So far, the military equipment has been rather expensive, and the receivers have not yet been bought in large quantities. With time, the costs will fall. But the big cost reductions await the arrival of mass markets. Millions of civilians will find Navstar to be very useful in their daily lives. When that happens, production rates will skyrocket, costs will fall, and prices will follow.

Since Navstar signals sent out by the satellites go everywhere, they are there for everyone's use—military and civil. Though the system is being put in place primarily for military purposes, I predict that we will soon see millions of civilian users tied to these satellites, because they offer a host of opportunities that have never existed before. We have not even begun to think of all the applications for its use. Navstar will be fully operational in the early nineties. At that time, people will begin to see how valuable an aid it is to many of their normal pursuits. As the market grows, costs will fall to the point where one can buy a receiver for several hundred dollars, depending on what features one wants.

For the car owner enamoured of the latest optional devices, and especially for the trucker who wants to avoid getting lost in unfamiliar parts of a city, Navstar makes possible an electronic moving map display showing with great precision the user's location. Moving map displays will probably not be a feature of inexpensive cars, but in BMW, Cadillac, Lincoln, and Mercedes automobiles they may very well become standard equipment. Thousands of navigational uses will become possible and practical because of the built-in precision of Navstar. For example, a messenger using a 9-digit zip code and a Navstar receiver will be able to locate obscure addresses in complicated cities like Boston or Los Angeles. The commercial sale of receivers and related location devices will add billions to the gross national product (either ours or Japan's). None of this would have been possible without Navstar or something like it. This is one of many new classes of commercial businesses that are enabled by space systems, no matter who puts the satellites in place.

Ships, boats, and airplanes—the more obvious users of Navstar—make up a market of about 1 million vehicles. Trucks might add another 10 million potential users. And if a large fraction of the 100 million cars in the United States were to become candidates for Navstar receivers, that would be a real mass market indeed!

So is navigation by Navstar a good example of space commerce? Perhaps not by the usual definition, but certainly this potentially significant addition to the gross national product could not take place without the existence of Navstar satellites. It is the same story that we saw with backyard dishes for TV reception. Many billions of dollars' worth of backyard antennas have been sold, even though some people would probably say they do not meet their strict accounting standards for space business.

The Soviets are putting up a system called Glonass (Global Navigation Satellite System) which will be very similar to Navstar. They certainly did not plan it for commercial use, since it predates Gorbachev's arrival, but Glonass satellites will no doubt be used by certain commercial customers in various places just as some people will use Navstar satellites.

■ GEOSTAR

As of 1990 the only private company to attempt to develop a navigation system is Geostar. This promising project belongs to a class of systems called Radiodetermination Satellite Service (RDSS). Four RDSS systems were allowed in rulings by the Federal Communications Commission in 1985 and 1986, but so far Geostar is the only entry in the field.

This system was invented by Dr. Gerry O'Neill of Princeton. When I first heard him describe it 8 years ago, it seemed neither practical nor very useful even if it worked. Over a period of time, O'Neill and his backers put some good engineering talent to work on it, and the system has evolved to the point of having great potential for a number of uses, one of the most interesting being as a truck tracker. Other moving objects like barges, railroad cars, and perhaps taxis are also potential candidates for service.

Geostar was originally billed as a great boon to air traffic control. As a former head of the Federal Aviation Administration, I was not impressed. What Geostar would have done, if it had been a practical design, was a small part of the overall air traffic control function. For the private airplane owner, it promised much but delivered little. And even if every airplane owner in the United States used Geostar's limited services, this still would not be a big enough market to allow Geostar to break even financially.

But with the help of engineers at both Comsat and RCA Astro, the present design eventually came into being, and organization of the company began in earnest. Former Secretary of the Treasury William Simon became board chairman, stock was sold, money was raised, engineers were hired, and systems tests were done. Geostar originally arranged to place its payloads on other people's geosynchronous satellites. This plan appeared to be a good way to save money, but it has not worked out well. The reason is that the particular satellites Geostar chose were either not launched into the proper orbits or failed partially in orbit. There are now two satellites in operation, but one of them has very limited life left. Geostar has talked about putting up expensive dedicated satellites at a cost of hundreds of millions of dollars. It may be able to raise the money, but more likely it will have to revert to the approach it began with—piggy-backing on other people's satellites.

For its initial stage of operation, Geostar planned to use two geosynchronous satellites in 24-hour orbits, in contrast with Navstar's 20 or so satellites in 12-hour orbits. Geostar would use ordinary but accurately timed radar-like pulses at K-band frequency to measure distance from receiver to satellite—not the extreme precision timing pulses of Navstar. Knowing the distances from the receiver to two widely separated satellites (whose positions are very accurately known), one can calculate the receiver's location on the earth's surface. (One also needs a topographic map of the area of interest.) The satellite sends a query signal that triggers a response from the receiver (transponder), thus enabling the required measurement of distance. In the normal mode of operation, the central control station polls a number of receivers to find their locations. Geostar claims that testing has confirmed the accuracy of such a system to be comparable with that of Navstar.

In early 1989 I visited Geostar headquarters in Washington, D.C., to see the system in operation. The antennas for the ground control station are located on the roof of a downtown office building. This central control station manages the system and creates a display or map of the United States. On the display I saw trucks that were scattered over the road system of the United States. The trucks were being queried as to their identity, and they responded with their locations—all of which were displayed in real time on a CRT screen. Such a system enables fleet management from a central location, since the same display of truck positions can be seen at the trucking company's own headquarters. Thousands of trucks were being monitored, with each trucking company seeing only its own trucks.

While the basic principles of a satellite-based RDSS system have been proven, Geostar does not operate in that mode because of problems with its satellites. The system using RDSS is now said to be the plan for Phase 2; under the present Phase 1, Geostar uses LORAN to determine the position of equipped vehicles.

Geostar should soon find out whether its services appeal to a broad market. An obviously important factor is price. Even though Geostar may offer an attractive service, it could easily price itself out of the market. It is anyone's guess how Geostar will fare in competing for truck tracking and other mass markets based on navigation and radio-determination satellites.

Geostar has not made money yet, but it has enough customers to show that there is a demand for that kind of service at an appropriate price. Both Navstar and Geostar have suffered from not having enough satellites in orbit, but this situation can change rapidly with time. Geostar's use of LORAN permitted it to begin to offer service and to demonstrate that its system will work. This put it in business on an interim basis. But it needs to control millions of trucks—not the thousands it had in mid-1990—to be profitable. Can it price its services low enough to attract millions of customers and high enough to make money? The jury is still out on this question.

In radiodetermination systems a central control station sends and receives messages to and from an equipped vehicle, calculates the position of the vehicle, and may send the location back to the vehicle or to another location. For 30 or 40 years, radio-dispatched taxis have used an earlier nonautomatic (manual) version of this technique. If Geostar had a taxi company as its customer, the taxi control station would have a terminal that displayed the locations of all its taxis and whether each had a customer. When a customer phoned to ask for a taxi, the controller could glance at the display and select a taxi to be sent to a certain nearby address. This instruction would appear on a small alpha-numeric display on the instrument panel of the taxi. The taxi would acknowledge the message by pressing a button. To make my example practical, the taxi driver would notify headquarters of his destination each time he picked up a passenger so they could estimate when he would finish his trip. From the information available to him, the dispatcher would call the taxi most likely to be available immediately and tell it to take care of the next customer.

Of course none of this calling back and forth needs to be by voice—it can all be done by very brief messages lasting only a thousandth of a second that are sent semi-automatically by black box. The real question is whether the taxi company can do enough extra business to justify the cost of the system. In some areas, radio congestion is so bad that taxis are not permitted to use the old method of endless conversation back and forth. The new system would solve many of those problems by eliminating long-winded conversations and substituting brief digital messages. These might occupy as little as one or two percent of the radio capacity used by voice-dispatching methods, and of course it operates in a different part of the frequency spectrum. Ostensibly, such conservation methods could be applied to any radio system.

One question faced by people offering radiodetermination services is whether their services are different in kind from communications services. They are being allocated a certain portion of the radio spectrum; they must justify themselves to keep title to the spectrum. If the position fixing appears to be only an ancillary part of the typical message they transmit, they may be in danger of losing their distinctive character and thus losing their frequency assignment. If, on the other hand, they do not allow their customers to communicate rather freely, some customers may want to shift to a communications service where there are no limits on traffic.

Whether this distinction is important will depend on how the market actually develops; my own view is that some customers will want to talk more than they can on the Geostar system, and they will gravitate toward systems that are mainly communications systems. Such customers will then use something like Navstar on those few vehicles that need navigation service but will equip most of their vehicles merely with a communications terminal. Other customers such as truck fleets may need to have position information on every vehicle but have no need for extensive communications with the trucks. They may be logical candidates for Geostar. Such users can get quite a bit of mileage out of Geostar by using canned messages. I doubt if there are more than a hundred things a controller will want to say to a truck driver. So for most purposes, he can use a numerical code of a few digits to cover almost all contingencies. But we should remember there is a shortage of bandwidth at L band.

▪ QUALCOMM

Qualcomm offers an option in Ku band, a frequency that is not so crowded, and uses a technique (spread spectrum) that does not require allocation of frequencies as long as the signal levels remain undetected by the primary users of the spectrum. Qualcomm seems to have come out of nowhere in a short time to take its place as Geostar's competitor. Qualcomm was started in 1985 by two people well known in communications—Dr. Irwin Jacobs and Dr. Andrew Viterbi. By 1988, they became operational with a position-reporting and two-way communications service. Since Geostar did not at the time offer two-way communications service, Qualcomm had a temporary advantage. Qualcomm's truck system, called OmniTracs, can display trucks' positions by means of an installed LORAN C receiver. The fleet owner receives messages from the data terminal in the truck that transmits the truck's location plus any message generated by the driver. Of course, the fleet owner might choose Navstar over LORAN, depending on accuracy requirements and costs. While LORAN is supposed to serve the whole country, there are places in the mountains of the United States where coverage is poor to nonexistent.

Qualcomm solved the problem of the lack of LORAN C in Europe by joining Eutelsat (see Chapter 2). Qualcomm has developed with Eutelsat what it calls QASPR (Automatic Satellite Position Reporting). It has demonstrated that two Eutelsats working together can locate the trucks in a scheme not very different from Geostar's ultimate plan. It claims 1,000-foot accuracy. Based on the use of existing satellites, it seems to have an advantage over Geostar, which requires special packages in the sky. Locstar, Geostar's European partner, plans to launch L-band satellite capacity to be ready by 1992. *Via Satellite*'s European correspondent, Pierre Langereux, said that Euteltracs service would be available in January 1991 using the QASPR technique.

Not only did Qualcomm come up with a technique obviating the need for use of LORAN in Europe, but the technique can be applied in the United States and elsewhere. Qualcomm has chosen partners in

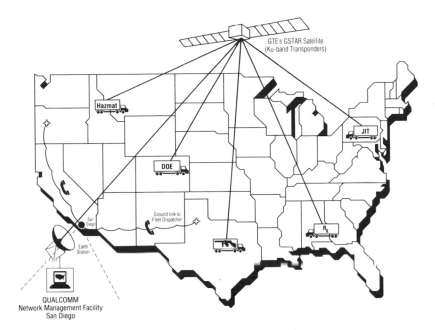

GTE's GSTAR Satellite
(Ku-band Transponders)

Hazmat

JIT

DOE

R$_x$

Ground link to
Fleet Dispatcher

San
Diego

Earth
Station

R$_x$

QUALCOMM
Network Management Facility
San Diego

Figure 16. Qualcomm's OmniTracs—a satellite communications system for truck fleets—provides two-way messaging and vehicle location for mobile units anywhere in the United States. Both the forward link to the truck and the return link are managed by Qualcomm's Network Management Facility in San Diego. Fleet data are then routed to the trucking company via terrestrial or satellite links. (Illustration courtesy of Qualcomm.)

Europe (Alcatel), Canada (Cancom), and Japan (Itoh) to offer equipment and service for trucktracking.

■ COMPARING NAVSTAR, GEOSTAR, AND QUALCOMM SERVICES

Navstar signals can be tapped at no cost to the user other than receiver cost. Geostar needs its own satellites, or at least payloads on someone else's satellites. Qualcomm is an intermediate case, paying someone for use of existing satellite capacity, but using it only two transponders at a time.

Navstar does suffer one serious disadvantage. Since it is basically a military system, its high-precision transmissions (at 10 meters' accuracy) are encoded in such a way that only those with proper military equipment can receive them. Civilian users must settle for the less precise signal (which the military says will always be of 70-meter accuracy or better) or equip themselves to make a differential GPS computation. This involves putting a GPS receiver and transmitter at a reference point of known location. The local transmitter can be used to improve the accuracy of Navstar to yield position errors of only a few meters. This obviously adds expense to the system and tends to drive the costs of using GPS signals in the direction of competing systems. But of course Navstar's coarse signals are as good as most customers will need.

Navstar's military character may also deter some non-U.S. users, who may wonder if its capabilities will always be available to them in the future. This prospect is especially worrisome now that the Air Force has become serious about encrypting Navstar's most accurate signals. With peace breaking out between the United States and the U.S.S.R., the Air Force will probably use the unsettled situation in the Middle East to justify its action. My suggestion to the Pentagon would be to announce that in any emergency short of an attack on the United States, Navstar's high-precision data will be available to all in good, clear form. This would encourage a great number of people to commit themselves to its use.

It is still too early to know how the civilian market will adapt to the availability of precision signals from either the moving Navstar satellites or the fixed Geostar satellites or other systems like Qualcomm's. Experience with the first-generation Transit system showed that services by such satellites were useful, even though one sometimes had to wait an hour or so to get a fix. Navstar has had some satellites in orbit for years, but only now is the full system being deployed. Consequently, the mass of potential users has bided its time, waiting to see that an operational system really exists. While Geostar and Qualcomm have both shown that there is a market for tracking trucks, neither has enough customers yet to make money.

I believe that we can expect interest in such systems to rise rapidly over the next 5 to 10 years. That is about how long it will take people to learn how well Navstar and other systems perform and to decide to go out and spend money. But that high rate of interest will come. The advent of high-precision navigation signals from satellites is likely to change forever our attitude about the advantage of knowing the exact location of a moving vehicle (or mountain backpacker, for that matter). It will bring equally impressive benefits to a host of users on nonmoving objects—for applications like surveying, mapping, charting, studying slow land movements, predicting earthquakes, and dozens more.

We frequently say that scientists and engineers always do a poor job of forecasting the future—they predict that more will happen in 5 to 10 years than actually does happen, and they predict that much less will happen in 20 to 50 years than actually happens. The new wave of progress in navigation and position fixing may be another case where, after a slow start, our use of such systems will expand extremely rapidly. This could turn out to be the next important space-based stimulant to the economy.

7

HABITATIONS IN SPACE

The number of opportunities in space expands each year as scientists, engineers, inventors, and entrepreneurs come up with new approaches to old problems by operating in space. But there are also new problems to solve in space, such as having to live and work for months or years in a very challenging environment. Designing and building habitations for use in space may be one of the most exciting and rewarding commercial space opportunities for the next two or three decades.

In 1985 the National Commission on Space held hearings around the country to allow our citizens to express their opinions on what the U.S. space program should encompass. The public hearings resulted in a large volume of testimony that gave us a broad perspective on what people think about space and its exploration. A view expressed by many was that space is a neat place to visit and maybe to live. In fact, several people recommended we make provisions for arthritics and others with similar afflictions to live in space. They assumed that such problems would be lessened if one did not have to fight against gravity all the time. Little did they appreciate that people left in space suffer not only from loss of muscle mass and bone calcium but a general malaise as their bodies atrophy. Space is probably the last place you would want to visit if you had a serious health problem.

But how can we create habitats in space where people can live, if not thrive, for extended periods? Typically, these habitats are designed to create what we call a shirtsleeve environment. This means people do not have to wear special clothing or masks but can move about, see what is going on, talk to one another, breathe normally, make extensive measurements, listen to music without giving thought to the fact that a few feet away is a hostile world. In the properly designed habitat, they do not need to spend a major fraction of their time and energy just staying alive.

To create that benign environment takes a lot of equipment. At first thought, we might conjure up a strong metallic shell that can withstand normal atmospheric pressure, like an airliner flying extra high, but we need much more. We must pack into the facility a thousand things, such as a highly redundant supply of air in pressure tanks, casks of water, storable food and drink, cooking facilities, filters to remove carbon dioxide and other impurities from the air, facilities for storage of feces, urine, and various kinds of garbage. For extended stays in space, we need a variety of foods, including fresh vegetables. We need hot and cold showers, recreational facilities, training programs, entertainment, and of course medical supplies and exercise facilities. Our first such habitat was Skylab.

▪ SKYLAB (MAY 17, 1973–JULY 11, 1979)

In astronaut Mike Collins's interesting book *Liftoff*, he describes the equipment and supplies carried on board Skylab, which could rightly be called the first space station. The astronauts had to keep track of some 12,000 separate items—and this was 15 years ago! Collins also pointed out the problem of finding one of those 12,000 items by searching through dozens of identical storage lockers, and the problems of putting away things that otherwise would float around the cabin, usually ending up partially blocking the air filter vents somewhere—and usually where they would be hard to find. Fortunately, Skylab had a good-sized room (an oxygen tank for Apollo) into which the crew tossed everything they thought they would never need again.

Although many people have already forgotten it, Skylab was one of the best-known experimental environments that has been put in space. It flew shortly after the end of the Apollo program, when NASA found it had several pieces of unused Apollo hardware left over after the moon program ended. They had bought enough to carry out 20 moon flights but cut the program to 17 after they saw that they had accomplished essentially all their objectives. Cutting the program short saved some operating costs but mainly saved hardware that thus became available for the Skylab program. NASA made a laboratory out of the third stage of a Saturn rocket that had been built to boost astronauts from earth's orbit into a moon trajectory. In 1973 NASA launched Skylab on a Saturn rocket and then launched three different three-man crews to Skylab using an early version of an Apollo rocket.

During the launch of Skylab, one of the two solar panels was ripped off by atmospheric drag, and the other failed to deploy until it was freed manually. (Skylab's failure to deploy its solar panels led to the first photograph ever taken by one satellite of another satellite in a different orbit. A photo-reconnaissance satellite sent back the data needed to confirm which tools the astronauts should take into orbit to free the stuck panel.) Despite these mishaps, Skylab was able to develop about two thirds of its expected power levels and do most of its planned experiments. It made history by keeping the first crew in space for 28 days, the second crew for 59 days, and the third crew for 84 days.

Skylab was a huge success as an experiment to measure many aspects of human physiology in the weightlessness of space. We and the Soviets—whose success in collecting valid data on the effects of space on human physiology has not been impressive—have drawn on it for years. Skylab also set new milestones in collecting earth resources data using multispectral cameras. These cameras later saw extensive use on Landsat. Its work on photographing the sun outside the earth's atmosphere was equally valuable. One of its crews recorded 10 miles of taped data similar to Landsat photos of the earth and also took 29,000 pictures of the sun. Mike Collins called Skylab not only a home away from home but a radio and TV station, a physics and biology laboratory, a doctor's office, a photo studio, and an astronomical observatory.

NASA planned to deal with Skylab's limited life expectancy in orbit by going up in the shuttle and attaching a rocket motor to it that could be used to extend its life by driving it to a higher orbit with less drag, or eventually to terminate it by using that rocket motor to slow it down enough that it would fall into the ocean far from land. The shuttle was delayed so long that it still had not made its first flight when Skylab fell to earth in 1979. (The shuttle first flew two years later.) For lack of a motor to control its reentry, Skylab disintegrated over Australia and got a bad name for littering the outback; NASA was heavily criticized for letting it fall on land where it could have caused severe damage (it didn't). But while it was alive, Skylab did excellent scientific work.

■ SPACE SHUTTLE

Following the last flight to Skylab in 1974 and the brief encounter with the Russians in the Apollo-Soyuz linkup in 1975, the U.S. entered a six-year drought with no astronauts operating in space. While the Russians took all the headlines, racking up one space feat after another, we focused on building our new manned vehicle, the space shuttle. The shuttle was conceived as part of the plan for manned space flight which followed the Apollo program. It was assumed that the next large project in space would be a space station—a permanent facility in space. As the Apollo program wound down, the Nixon Administration debated whether there was sufficient justification for starting such a program, recognizing that it would take many years to complete and would cost billions. It was decided not to embark on such a large and expensive effort. The space shuttle, on the other hand, was of more modest scope and would be needed to get to the station in any case. If properly designed, it would have a lot of capability on its own and would be a good beginning. The shuttle was thus a compromise between beginning the development of the space station itself and putting the whole question off indefinitely.

To gain more support for the space shuttle, it was sold as the Space Transportation System—the STS—which would be the cheap way to get to orbit. With a reusable vehicle, it would make space flight a routine operation. Enthusiasts said it would soon be flying about once a week into orbit beginning in the late seventies. When it finally flew in 1981, it had become the only means for getting into space. The decision to phase out all other means had been made because certain officials felt it was the only way to avoid cancellation of the shuttle during the Carter years. The shuttle's lateness and large cost growth—and the government's unwillingness to 'fess up to its true cost and complexity—had helped deflect the public's interest and enthusiasm from the space program, including the shuttle.

Once it had flown successfully and demonstrated that it did indeed have the ability to launch satellites, the space shuttle embarked on a 5-year period of increasingly demanding flight rates. While not a cheap way to orbit, it did have some desirable features that rockets did not: it could, for example, recover satellites and bring them back for repairs. This was a totally new capability which many came to appreciate when the shuttle recovered one NASA satellite and two commercial satellites from orbit. It also rendezvoused with a fourth satellite which had malfunctioned and allowed the shuttle crew to make necessary repairs to get the satellite on its way to geosynchronous orbit.

Seeing that the shuttle could retrieve satellites made the Russians very nervous; they imagined us going up and examining their secret satellites without their permission or even bringing them back for analysis. That feature became their excuse for not being willing to negotiate seriously toward an antisatellite treaty. For 5 years things seemed to be going well for the shuttle. Then came the *Challenger* accident, which put the shuttle out of business from January 1986 to October 1988.

The accident was a very serious and traumatic event for the nation. It shattered morale as well as many reputations at NASA and caused much soul-searching about NASA's role and the shuttle's place in NASA. In the aftermath of the *Challenger* accident, there is still a difference of opinion within NASA on how often the shuttle should fly. One school wants to treat it as a scarce resource and fly only when necessary, thus reducing the probability of another accident; the other school wants to fly as often as possible to reduce the cost of each flight to the minimum. A rate of 10 to 12 flights per year seems reasonably achievable at this time.

I hope we have learned that the shuttle orbiters should be treated like commercial air fleets. All airlines know they must plan for replacement airplanes on a regular basis. The shuttle is no different; we should assume that one or more orbiters will be damaged beyond repair within the next few years. While we cannot say when such accidents will happen, we know that as time goes on, they are almost a dead certainty.

The orbiter is configured somewhat like a cross-country truck; the cab of the truck has room up front for the crew and room behind the seat for a spare driver to sleep. The shuttle has room for the flight crew in the cockpit during takeoff and landing; adjoining the cockpit is enough room to eat, sleep, and do limited work and exercise. This space is the size of a small room. The rest of the shuttle's useful volume is the unpressurized payload bay behind the crew compartment; it is comparable to the truck's trailer and is even larger than most truck trailers, measuring 15 feet in diameter by 60 feet long.

Much of the shuttle's work so far has been launching satellites. Depending on the size of the satellites, several may be carried aloft at once and launched sequentially from the payload bay. Various satellites have been launched from this bay and a few have been retrieved and brought back to earth, stored in its capacious maw. Experiments to be performed on the shuttle normally must be confined to the crew compartment—a rather small amount of space, and not very effective as a laboratory. Means were developed to carry unattended experiments in the payload bay on pallets, and a number of such pallets have been carried aloft. If we wanted more space for human-tended experiments, we would have to build a closed facility which could be carried in the payload bay. Several versions of possible facilities were studied—some of which would remain in the shuttle payload bay and some of which might venture forth into space on their own.

■ SPACELAB

While considering ways to extend the operational capability of the shuttle, we engaged the Europeans in discussions about their possible participation in the manned space flight program. Before the European Space Agency (ESA) was created, the European Launch Development Organization (ELDO) suggested to NASA that an upper stage to take shuttle payloads to geosynchronous orbit could be their contribution to a joint space program. NASA was not enamored of the idea, having some plans of its own for meeting that need. Eventually, both McDonnell Douglas and Orbital Sciences built commercial versions of the upper stage for the shuttle. McDonnell Douglas calls its version PAM; OSC calls its version TOS (see Chapter 4).

In the process of considering the pros and cons of the European proposal, NASA came up with the proposition that its proposed "Sortie Can" could be taken over by Europe. The Sortie Can was not a well-developed concept, but NASA knew that we had to have a much larger laboratory than could be crammed into the living quarters in the forward compartment of the shuttle. (This compartment had been shortened even more to accommodate longer Air Force payloads in the shuttle payload bay, thus compromising further the experiments that could be done in it.) Discussions with ESA led to the decision for Europe to construct Spacelab, which could be looked on as the full-fledged version of the Sortie Can.

Spacelab essentially converts the payload bay into a laboratory, half being enclosed and half unpressurized. Europe saw merit in using Spacelab as a workplace in space for their own scientists working alone; furthermore, NASA agreed to buy from Europe any additional Spacelabs it might need. Although NASA had never committed itself to buying a given number of Spacelabs, it seemed that at least 3 or 4 and perhaps as many as 8 would be needed to carry out all the flights that were being talked about at the time. NASA planned to fly the shuttle about once a week, and a significant fraction of those flights were expected to be Spacelab flights. Several years of consultation led to a design of Spacelab which the Europeans agreed to pay for and supply to us. They would use the first flight for their own purposes; after that we could decide who would use it and how and when. ESA funded the construction of the first unit, with the understanding that any subsequent Spacelabs would be bought by the United States from Europe.

Spacelab is a cylindrical laboratory about 24 feet long and 14 feet in diameter. When the shuttle carries Spacelab into orbit, the space available for the crew to do lab work is extended significantly—by something like a factor of 6. Spacelab has been flown three times so far and has conducted many useful experiments. Spacelab comes in two forms: a habitable module which is used as a lab wherein people can work in a shirt-sleeve environment, and a pallet open to the outside environment which holds various experiments that are remotely controlled from the shuttle cockpit. These two components of Spacelab can be flown together or separately. Because Spacelab is attached firmly to the shuttle when it flies, if the shuttle stays up a maximum of 7 days, that becomes the maximum stay of Spacelab, even though it is capable of 30 days in orbit.

So far, we have chosen to buy only one Spacelab from Europe, mainly because of the greatly reduced shuttle flying rate compared with what was planned in the beginning. Under the present plan of about 12 shuttle flights per year, Spacelab will fly perhaps twice per year. This means the original understanding about the number of Spacelabs the United States would buy from Europe is totally obsolete. From the European standpoint, the big investment they made in

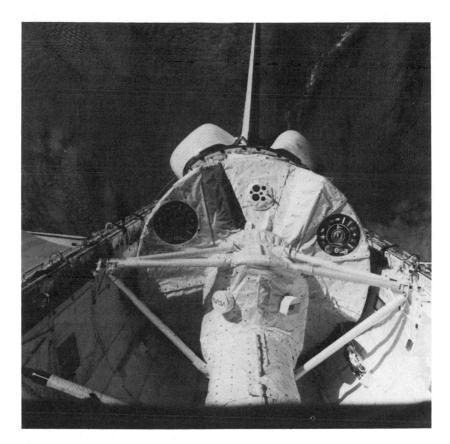

Figure 17. A handheld Hasselblad camera was aimed through the aft windows on the flight deck for this scene of the active Spacelab module in the cargo bay of the earth-orbiting space shuttle *Columbia*. The docking tunnel, leading from the shirt-sleeve environment of Spacelab, is in the foreground. (Photograph courtesy of NASA.)

Spacelab would have been impossible to justify if the United States had planned to buy only one Spacelab. At a conference in Hamburg about 5 years ago, a high European official said cryptically that the United States is due only one major gift per century from Europe. Last century, it was the Statue of Liberty; this century, it is Spacelab.

Spacelab is a very sophisticated laboratory capable of doing major work for its operators—European, American, or otherwise. It has two serious disadvantages: one is that it almost monopolizes the shuttle. When Spacelab is on board, the shuttle can do very little other work—whether on the way up, in orbit, or on the way down; its payload bay is occupied fulltime by Spacelab. The second is the long leadtime necessary to set up and execute experiments. Spacelab will be very good for materials processing that needs manual attention. Although some people have assumed that materials processing research will proceed better without people around to cause vibration, the very nature of lab work is that we do not know all we need to know about the process. Therefore, human monitoring and intervention is likely to be necessary intermittently throughout the experiment. Because of this, we recognize that Spacelab is a very important asset and no doubt will remain so for many years to come.

From the standpoint of the United States, Spacelab is not a commercial facility; nonetheless, the Europeans thought it had commercial possibilities, in that they expected to sell several of them to NASA for our own lab work. But in Spacelab, the United States enjoys several of the advantages of having a commercial space program. One advantage of such a program is that space capabilities can be created without investing taxpayers' money; in that sense, Spacelab is commercial because, as OMB has noted, it was designed and built with someone else's money. ESA spent $1 billion on the first one; we bought the second for about $500 million. Can we get other facilities without using taxpayers' money? Are there good commercial opportunities in providing space habitats? The answer to both questions is certainly yes—as we will develop in later sections of this chapter.

■ EXTENDED DURATION ORBITER

One of my pet ideas, during all the years that the shuttle was being developed and for several years thereafter, was that we should extend the duration of the orbiter's time in orbit to about a month. The present design provides for only a week in orbit, the exact time depending on which chores the crew must carry out before returning home. I made numerous speeches (with no visible result) on how that staytime should be gradually increased—first to 2 weeks, then to 3 and 4 weeks—to allow us to collect significantly more data on microgravity and other space-related experiments than we could get on the short flights the shuttle was designed for.

Doubling the length of the flight would more than double its utility because a lot of the present time in orbit is taken up either getting ready to work in space or getting ready to come home. We could collect much more data at only slightly more cost if we stayed in orbit longer, because most of the cost relates to getting up there in the first place. Depending on how one calculates the numbers, it costs about $200 million to make a shuttle flight. This means it costs about $30 million per day to operate the orbiter in space. If one doubled the length of the flight, one could cut costs of productive time to less than $15 million per day. The equipment needed to stay in orbit a few more days is not expensive and—if installed—would markedly reduce the average cost per day. NASA estimates the cost of modifying an Orbiter to permit staying in orbit several weeks at about $200 million—an amount that would be recovered in a very few flights of longer duration. Although it is not known whether any of the speeches I and other similarly inclined people made on the subject had any effect, I am happy that NASA decided about a year ago to begin work on the concept. The program is EDO—the Extended Duration Orbiter—which begins by extending flight duration to 16 days, with an option of going to 28 days at a later time. The program increases storage capacity for food, oxygen, fuel, and waste disposal to permit the longer stays.

EDO might be considered a commercial program, depending on one's definition of the term. Owen Garriott, a retired Skylab and Spacelab astronaut who appreciated the value of long stays in orbit, probably had more to do with the concept's being sold than anyone else. As an officer of Teledyne Brown, he proposed that his company take on the job of extending the duration of the orbiter's staytime in space. The company would modify an orbiter at no cost to NASA; it would get its money back by collecting a fee for each day the orbiter flew beyond its normal mission length.

When NASA bought the concept, it also decided that it would be difficult for any contractor other than Rockwell—the original builder of the orbiters—to do a creditable job of EDO. NASA gave Rockwell the contract to supply an EDO package for the orbiter *Columbia* at a price of $53 million. NASA is not making progress payments, as is customary in such cases; it has asked Rockwell to carry all the costs until EDO is delivered. At that time, NASA will pay 1/3 of $53 million, with each of the other two payments to be made at intervals of one year. The cost of the accumulated interest for the delayed payments is estimated at $14 million, using nominal interest rates. Rockwell is to recover this investment by having commercial customers pay $1.8 million per day of extended time on any flights where they are permitted to fly extra days in orbit. If NASA chooses not to permit any commercial customers to fly for extended times on *Columbia,* then Rockwell will have to absorb the loss, but it is not expected that NASA will be unreasonable.

I believe that the most significant payoff for EDO will be missions that carry Spacelab. One might argue that the flight deck on the shuttle is too small to ask crews to do useful work for a month, with all the possibilities for short tempers engendered by such cramped quarters. Spacelab can carry much more sophisticated equipment and offers many more opportunities for crew/equipment interaction than the shuttle flight deck does.

Although EDO is not a very good example of the private sector assisting in carrying out NASA's program, it seems to be the first instance where NASA has opted to introduce a commercial factor into shuttle operations. This may be a good example of the problems the private sector has in commercializing chunks of the NASA program. In this case, Teledyne Brown spent several years and probably several million dollars designing the EDO system and trying to sell it to NASA and to Capitol Hill. In the end, NASA benefited, so the public will also benefit. All we can say to those who pushed the idea and lost money is: thank you, and better luck next time!

From time to time discussions have been held on supplying commercial power to the shuttle so that its flight duration time could be increased. General Space Corporation believed that it could build power packages that would remain in orbit continuously, allowing the orbiter to fly up and plug in at will. Other schemes used collapsible panels of solar cells that could be extended once the shuttle reached orbit; these would pick up from the sun perhaps twice the power of the basic shuttle power supply. Under a contract with Able Engineering such a panel was deployed experimentally from the shuttle, and it was said to be a very efficient package—storable in a very small volume when

Figure 18. Artist's concept of the enhanced configuration of Space Station *Freedom*. The mission of this permanently manned international space station developed by the United States and its partners—Canada, Japan, and 9 European nations—is to enable humans to live and work in space for extended periods of time and to make full-time observations of space and the earth. As an orbiting science laboratory it will allow investigators to conduct fundamental research in materials processing and the life sciences. (Illustration by Alan Chinchar; photograph courtesy of NASA.)

not in use. But so far, none of these tentative proposals has resulted in a commercial activity.

■ SPACE STATION *FREEDOM*

In the mid-1990s assembly in orbit of the Space Station *Freedom* (SSF) should begin, with about 3 years elapsing before it goes into full service. The station is basically a NASA-owned and operated unit, with several international partners supplying additional working space and equipment. Its heart will consist of 4 main modules, two for the United States and one each for Japan and Europe. The working space in the station initially will be those 4 modules—each 15 feet in diameter and 40 feet long—plus the so-called nodes. The nodes contain ports or hatches that allow the shuttle and other spacecraft to dock at the station and also serve to connect the modules together. The nodes are attached to a cylinder about 20 feet long that bridges the gap from one module to the other; the whole node enclosure is large enough to contain much equipment.

Each module will be carried up as a full payload-bay package in the shuttle. About 20 shuttle flights will be required to carry up all the main components of the station—the modules and nodes, the structural elements that make up the main beam on which all components are mounted, the power-generating solar panels, tanks for fuel and other storable items, communications systems and antennas, plus experimental equipment that will be externally mounted, such as cameras and telescopes.

The United States will supply a laboratory module and a habitat for the crew—all 8 crew members being housed in the U.S. module. NASA is also supplying all support equipment, supplies, and basic infrastructure—power, food, water, and so on. The Japanese and European modules will be laboratories designed for their own experiments, but at least half of usable time on all modules will be available to American crews. Joint activities are planned so that there will be opportunities for some of the crew to operate in a proprietary mode and also in a mode where all work together on experiments of interest to all parties. Other laboratory modules are already under discussion; one example is a centrifuge suitable for experiments with people. The original centrifuge, which is expected to be housed in a node, is marginal at best for testing animals as large as people. In time, a centrifuge several times the diameter of the first one (which is limited by the 15-foot diameter of the shuttle) will be required and will be placed in orbit near the station, probably flying in formation with it.

Canada is supplying the remote manipulator arm. This tele-operated arm allows an operator to sit inside the orbiter and handle a spacecraft outside, positioning it to suit whatever experiments or operations may be going on at the time. Altogether, the portions of the station supplied by non-U.S. entities will cost about $2 billion.

While there has been much talk of the private sector furnishing certain components, facilities, and supplies or assisting in the construction or operation of the station, so far nothing of the sort has become firm.

It was proposed, for example, that certain robotic components could be commercialized, but this has not developed. Boeing made a specific proposal to finance part of the order it received from NASA for space-station hardware. In addition to modules covered under Boeing's billion-dollar NASA contract, the company's proposal would offer $140 million worth of extra logistical modules under a lease arrangement.

The problem with most such ideas is that they involve a private entity investing money to develop and build the desired item, with the cost to be recovered over a period of many years. This means that the cost of money will be a large fraction of the total cost. It eventually comes down to how much the company must charge for use of its property in order to recoup the money invested and all the interest paid before the company recoups its investment. Since the government can borrow more cheaply than the company, when the government looks at proposals of this type it usually decides it is cheaper to buy the item than to go to all the trouble of having the company supply it. That was in fact the decision NASA made about the Boeing offer; in May 1990 it gave Boeing the obvious response to its 20-month-old proposal.

The Space Station is officially expected to have a 30-year life, but it should last indefinitely. We are not sure whether it will be damaged seriously by meteorites, but it will be protected from all except the very largest of them by a shield. The chance that a large object will strike the station is remote, though not impossible; therefore, we must be prepared to repair the damage.

NASA recently received some valuable first-hand information on what space debris and meteoric dust are encountered by typical spacecraft. NASA's Long Duration Exposure Facility spent 6 years in orbit and was recovered in early 1990 by the shuttle. Its surface was covered by test panels of various materials to see how well they stood up under radiation exposure and physical collisions. There was no evidence of encounters with large particles, but most of the outer surface was pockmarked by strikes of infinitesimal dust particles.

■ THE SHUTTLE'S EXTERNAL TANKS

Once we have extended the shuttle flight time as much as seems prudent using the EDO concept, and once the Space Station is in orbit, we can then think about other ways of making long-duration flights, both manned and unmanned. One scenario I have had in mind since the early days of planning for the shuttle was to put a few of the shuttle's external fuel tanks in orbit. The external tank contains roughly 6 times the volume of the orbiter's payload bay. Even the tank's so-called intertank (the space between the hydrogen and oxygen tanks) contains about half the orbiter's volume. The external tank could be some of the cheapest real estate we could put in orbit. When it is jettisoned and allowed to fall back into either the Indian or the Pacific Ocean, it is already at about 99 percent of the velocity needed to go into orbit. This means we are losing a very valuable asset when we cast it off and let it fall into the water, because we invested a lot of money in giving it all that velocity.

Ten years ago, as we neared the time when the shuttle would fly, I thought it would make a lot of sense to put up some tanks to see how useful they would be. We could have put a few tanks in orbit and used the orbiter as a base from which to go out to convert the tanks into laboratories and habitats. Then we could have taken up equipment and/or crews for certain basic experiments. Much might have been learned about how to design the Space Station—and maybe much money saved—if we had experimented on a small scale before going straight to the Space Station design with no intermediate step.

Mine was not the only voice asking for the external tank to be put in orbit; many people spoke up during the years when the shuttle was being developed, but somehow those in charge were too preoccupied with other problems to pay serious attention to the possibility. Or perhaps officials at NASA, having been stung with criticism for allowing Skylab to fall out of control, were reluctant to put even larger structures into permanent orbit at that time, running the risk of losing control of them.

External Tanks Corporation (ETCO) is trying to make a business of putting some of those tanks into orbit. Another company with similar ideas is Global Outposts. Since many of the external tanks still have large amounts of residual fuel on board when they are jettisoned, it would be very easy to boost those tanks all the way to orbital velocity. There are two ways of doing so: (1) leave the tank attached to the orbiter until the whole assembly reaches orbital velocity, or (2) mount a rocket motor on the tank which will burn enough fuel to achieve orbital velocity. Of course the rocket must be controllable in direction—both to get the tank in the proper orbit and to deorbit it at the desired time—so the installation would be more than just a motor.

The TRW Corporation's orbital maneuvering vehicle (OMV) would be an ideal means for maneuvering the tank into position once it had been put into orbit—preferably by leaving the tank attached to the orbiter all the way to orbit. One may ask whether the manifest for the shuttle leaves enough fuel on board to take the tank to orbit; the answer is yes for some flights and no for others. Some have excess fuel left over; some are very heavily committed to the payload in the payload bay. For those shuttle flights without enough fuel to put the tank into orbit, the answer is simple: let the tank fall back into the atmosphere and wait for the next flight.

If put in orbit, the tanks could be an excellent laboratory for microgravity work, a storage facility, and a warehouse for things needing repair, a gymnasium, a fuel depot and fuel conversion station, a water recycling and purification facility, a repair and service station, a farm for growing fresh vegetables, a remodeling and refurbishment station, a manufacturing facility, or even an extra-large trash can. With people becoming more conscious of the need to clean up space debris, this last possibility is not outlandish. Colliding with garbage in space—especially an old rocket motor moving at several thousand miles per hour—is dangerous to your health. Smaller pieces of junk, being much more numerous, are probably even more dangerous.

ETCO first negotiated an agreement with NASA that would turn 5 tanks over to ETCO after they were jettisoned by the shuttle crew, with the tanks to be used for certain experiments before falling back into the atmosphere. Then a second agreement was negotiated whereby NASA would turn 5 tanks over to ETCO in orbit. (NASA reserved the right to certify the safety of the modified tanks, of course, and to deny flight opportunities if not satisfied.) All together, 10 tanks are available to ETCO.

One area of research where the tank may actually be better than the Space Station is in studying the effects of artificial gravity. Although we may not need to simulate 1 g, it is probably impossible to do so within the confines of the current Space Station design; it is just too small. With a diameter of 15 feet, the largest centrifuge you can squeeze into the Space Station as now configured has a radius of about 6 feet. It is impossible to envisage getting useful results by putting a person in such a device. Such rapid spinning would upset the basic measurements on the effects of artificial g. The external tank, on the other hand, is large enough for a centrifuge of 12-foot radius. While still not ideal, this is much more workable.

Supporters of the existing design of the Space Station say that the current design is adequate for working with rodents; this may be true. They say we might even be able to do useful work on primates, but this is more questionable. Maybe work on medium-size animals is practical; for primates and people or other large animals, we probably need a free flyer that orbits near the Space Station, with movement back and forth between the two. The external tank could be the free flyer with a reasonably large centrifuge inside it.

The limiting factor on what constitutes a workable centrifuge is rotation rate. Many experiments can be run at fairly high rates, but if we want to isolate the effect of rotation from that of the desired linear acceleration, then very small centrifuges are inadequate. In fact, for many purposes, we need centrifuges that can produce the necessary level of g at rotation rates limited to 1 rpm.

Accurately controlled conditions are needed for research, but if we want artificial gravity merely for the purpose of maintaining health, less than ideal conditions might suffice. One way to expose astronauts to artificial g would be to set up a circular track inside the tank; a person running around the tank on the inside would experience artificial gravity of about $\frac{1}{5}$ g. This might be enough to keep a person in reasonably good physical shape; certainly it would beat going for months with no exposure to gravity at all. Of course the artificial gravity would disappear as soon as he/she stopped running. Some have suggested putting astronauts in centrifuges where they would sleep, read, or do other work that does not require them to be oriented in any particular way to the workplace; that way they could be exposed to some fraction of 1 g for extended periods.

We may very well decide to rig up two platforms tied to each end of a tether so that we can spin the whole thing, thus achieving a large fraction of 1 g at a very slow rotation rate. Two external tanks tied together and rotating about a common center of gravity would make an excellent lab for testing the effects of artificial gravity of different levels.

ETCO is planning a set of activities for the orbital mode and the suborbital mode, with NASA to do oversight for safety reasons. We can easily dream up a long list of possible uses for tanks in orbit in addition to those listed just above. Here are a few ideas (partly redundant with items mentioned earlier).

Suborbital tanks

1. Thermosphere characterization, results telemetered out.
2. Microgravity crystal growth, results telemetered/parachuted out.

Orbital tanks

1. SSF has no gravity; spin up a tank about its axis.
2. SSF is too small for centrifuge; put centrifuge in a tank.
3. SSF crews need long baseline for lab test: do it in a tank.
4. SSF crews fighting over energy; collect more on tank surface.
5. SSF power shortage at night; store power in tank fuel cell.
6. SSF needs water and oxygen; process/recycle them in a tank.
7. SSF needs fuel for drag makeup; get it from a tank.
8. SSF crews need fresh food; grow it in a tank farm.
9. SSF machinery breaks down; fix it in a tank.
10. SSF crew wants to weld toxic material; do it in a (vented) tank.
11. SSF is crowded; move half of excess to a tank.
12. SSF crews go stale and atrophy; put a gymnasium in a tank.
13. SSF crews get claustrophobia; let them run free in a tank.
14. SSF crews become antisocial; banish them to a tank.
15. SSF generates waste; process (or hold) it in a tank.

While it may be years before anyone needs such a large volume for a space factory, in the meantime there is much research and development to be done where the tanks would be a reasonable size. Martin Marietta, the company that builds the tanks, did extensive planning for using tanks as labs a few years ago. One such experiment would use the tank as a gamma ray observatory. This experiment was large enough to justify a sizable unmanned facility in space which, with its long baseline, could become a "telescope" to determine the source or at least the direction of arrival of gamma rays from outer space. The tank met the conditions very well. The obvious agency to conduct the tests was NASA. What Martin Marietta hoped to do was to configure a tank as a lab before it was launched, then fill it with fuel. When the fuel was expended, and after a time for the tanks to dry out, experimental work could begin. For doing all the work to make the tank a lab, they could get extra income from NASA. While NASA found the proposal interesting, it was never interesting enough to attract any of NASA's money.

While there is some doubt that NASA would ever agree that a private company could take a tank, fill it with the necessary test equipment, run an experiment and collect data, turn the data over to NASA, and be paid for the results, nonetheless this kind of thing was contemplated in the President's space policy statement of February 1988. That statement included provision for making the external tanks available for commercialization, but it did not go into detail. Setting up such a facility using a tank that had served its purpose and been discarded by NASA may meet one definition of space commerce.

ETCO's plan has been to find private agencies wanting to do research in space. One of ETCO's stockholders is the University Corporation for Atmospheric Research (UCAR). UCAR as stockholder makes sense, since UCAR is a consortium of some 58 universities, all of which are research-oriented. Although UCAR means atmospheric—not space—research to them, many of these universities have faculty members interested in space research, making them natural allies in bringing the tanks on line for such purposes.

Other potential customers for tanks are foreign governments looking for labs in space. The purpose of making the tanks available for commercialization was not so that some company could take the tanks and offer them back to NASA; the purpose was for companies to find entirely new uses for the tanks. Nonetheless, it has been shown dozens of times that private customers—including foreign governments—are much more likely to sign up to use certain facilities or services if they have already been bought by a U.S. agency such as NASA. Use by NASA in such a case lends an air of authenticity to the concept, and the fact that it was known that NASA had done a safety audit, for example, could be very important to the willingness of a potential customer to sign on. The example immediately following makes this point very well.

■ INDUSTRIAL SPACE FACILITY

A variation on the theme of Spacelab is the Industrial Space Facility, which is a private undertaking by Space Industries, Inc. (SII). This company was started by Dr. Max Faget, a NASA retiree who decided that we needed something similar to Spacelab that would be human-tended as opposed to manned. "Human-tended" means that the facility would go up in the shuttle, would be parked in the shuttle orbit, and would be visited by astronauts from time to time. They would set up one or more laboratory experiments that would proceed on their own for periods of weeks or months. Then a shuttle crew would return to the lab, either to fix things that had broken or to change the equipment and get ready for another experiment. By not having people on board continuously, ISF would provide a more benign setting for certain experimental work such as growing crystals. The absence of people would greatly reduce the problems of disruption, contamination, and vibration. When necessary, the ISF could be retrieved and brought back from space. Otherwise, the results could be brought back by astronauts when "tending" the facility.

Dr. Faget has pursued the concept for 8 years and at various times has promoted it as a commercial venture. The thought was that NASA might rent some time on it, but SII would find various customers in the United States, Europe, and Japan, so that ISF's costs would be shared by many users. Thus NASA would get access to the facility for a fraction of its total cost. After Faget failed to find enough customers to make it a commercial enterprise, he approached NASA to see if NASA would rent it fulltime from him. At one point, he proposed that NASA lease it for 80 percent of the time, giving him the opportunity to continue looking for commercial customers.

In 1987 OMB decided that the Industrial Space Facility was such a good idea that it ought to be a candidate for an almost fulltime longterm lease by NASA, and some $25 million was put in the NASA budget to begin the process. NASA expressed interest in using the ISF but eventually decided it could not justify a sole-source procurement. Instead of leasing time on the ISF, NASA drew up specifications for a generic ISF to be called the Commercially Developed Space Facility (CDSF). With Congress looking over its shoulder, NASA asked the National Research Council (NRC) of the National Academy of Sciences to investigate the extent to which the CDSF would have scientific utility. It also asked the National Academy of Public Administration to price CDSF, assuming NASA's lease would probably have to cover the total cost of the facility.

The NRC's report, released in 1989, confirmed that various facilities are needed in space for materials research, but questioned whether

CDSF would be needed any time soon, given that we have not established the need for a production facility in space, only a research and development facility. For R&D, we have the present orbiter and anticipate having the "stretch" orbiter, good for 16 days initially and perhaps 28 days, followed in the mid-90s by the Space Station. Even though CDSF seemed to be a good place to produce materials in space, the NRC argued that we had not yet determined whether such work was important. In other words, we had not yet done the basic research and development to establish both the need for production and how to do it.

Moreover, for research (as opposed to production), we would need manned facilities, and the CDSF was only "tended." How could we do R&D if there were no people present to recognize when or if something unexpected had happened? No one would be there to fix anything that might break, or to note what did not go the expected way. Unmanned research and development was ruled out for the near term, but it was stated that some day, if developments in automation and robotics progressed well, we should be able to do tended R&D. When that time came—if it ever did—the CDSF could become very valuable, according to the NRC report. In the meantime, we should use the shuttle, the stretch shuttle, Spacelab, and perhaps Spacehab—a system described below.

The National Academy of Public Administration (NAPA), which finished its study at about the same time that the NRC released its report, said that the CDSF would be very expensive and that Space Industries' estimate of the cost of the Industrial Space Facility (some $700 million) was no doubt too low. They said that the actual costs to NASA, taking into account the rules of the game which called for a competitive procurement, the cost of money, and the rates of return necessary to attract capital, would result in NASA's having to pay more like $2.5 billion over a five-year period. NAPA implied we should not spend that amount of money without a very compelling need. (If we paid ESA $500 million for Spacelab, the estimate for CDSF may well have been too low, but Max Faget said he could build it more cheaply using private initiative and highly motivated people. Perhaps so.)

I am not optimistic about the CDSF's becoming a viable venture because of all the baggage that has been heaped on it by the NRC report. Certainly we cannot expect any action on it any time soon.

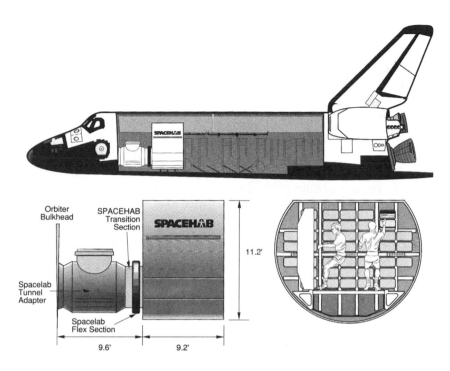

Figure 19. Spacehab, a privately developed pressurized 10 X 13 foot module, is designed to double the existing habitation volume of the space shuttle orbiter's middeck. It will also quadruple the volume available for human-tended space experimentation, while still leaving room for other payloads in the rear three quarters of the cargo bay. The first Spacehab flight is scheduled for 1992. (Illustration courtesy of NASA.)

Even taking an optimistic view, the market will probably not support more than one such habitat in the foreseeable future; this means that the development costs of such a facility could not be prorated over a number of units.

■ SPACEHAB

Spacehab, on the other hand, is a type of facility that is certainly useful and may be affordable. A concept of Spacehab, Inc., it is essentially a shortened version of Spacelab. Whereas Spacelab takes up most of the shuttle bay, Spacehab takes up only about one fourth of it. This leaves room to carry other payloads into orbit—for example, satellites to be launched. So there will probably be many shuttle flights that can combine Spacehab with other missions. It remains in the shuttle bay during the flight and hence does not need to incorporate all the functions that the Commercially Developed Space Facility must have to operate autonomously. For all these reasons, its proponents say that it costs less than CDSF by a factor of about 5 or 6. This is an important distinction; whereas rounding up enough money to launch CDSF ($700 million—or, if we believe the NAPA study, much more) is going to be very difficult, it is not unreasonable to expect to get, say, $125 million from investors to cover the cost of Spacehab.

As for the cost of getting such facilities into orbit, NASA has been willing to negotiate terms with organizations like Space Industries, Inc., and Spacehab under a fly-now-pay-later plan. At various times NASA has offered "space systems development agreements" (SSDAs) to startup companies, with the actual terms depending on specifics of the flight plans but also varying with NASA's desire to be helpful in getting such companies off the ground. Under SSDAs, NASA may offer easy terms, extending for up to 5 years the repayment of expenses incurred in putting payloads into orbit. If a company that had been granted an SSDA failed to attract customers, then NASA would have gambled and lost. But for SSDAs involving more than one flight, NASA would not continue providing free flights to a space venture that failed to produce users willing to pay the cost of flying. NASA has also offered "joint endeavor agreements" (JEAs)—where the company puts up the cost of the experiment and NASA puts up the cost of the flight. Later on, if the product is a commercial success, NASA expects to be repaid for those deferred costs.

James Beggs, a former administrator of NASA who is now chairman of Spacehab, believes that there are many customers abroad as well as in the United States waiting to use Spacehab's facilities and capabilities, so he is optimistic about Spacehab's future. Spacehab has an SSDA agreement with NASA that requires making progress payments over a period of years while Spacehab is establishing itself. After the first successful flight, Spacehab must pay the full cost of the flight. In 1989 NASA decided that Spacehab was a good idea but it would have to be procured through the normal competitive process. A competition was held in which Spacehab turned out to be the only bidder. The status of Spacehab in mid-1990 was that financing appeared to be available once NASA had produced a final certification of need, but as of this writing NASA is still bogged down in paperwork.

This is a case where the NASA imprimatur seems essential; several foreign governments and companies have vowed they will become customers on Spacehab once NASA signs up for time. Negotiating with them will be interesting; Spacehab will need to receive enough money per flight to amortize its costs over a half dozen or so flights and also pay NASA for its transportation costs. This probably means a flight must earn about $40–50 million to pay off. In any case, all present signs are that it will fly.

■ SPACE HABITATS FROM OTHER NATIONS

The Europeans have experimented with small unmanned free flyers for some time, using our shuttle to carry them into orbit, set them free in space, and retrieve them later. In the early 1980s the German company MBB-ERNO built a small free flyer called the Shuttle Pallet Satellite (SPAS) that carried a number of optical and other experiments. Now the same company is building the European Retrievable Carrier module (EURECA) under a European Space Agency contract. EURECA is expected to be deployed by the shuttle in the early nineties, to spend 6 months in orbit, and to be recovered by the shuttle. It can carry about 1,000 kilograms of payload in a volume of about 10 cubic meters. It should be able to make about one flight per year. (An American version of EURECA is the American Microgravity Carrier—AMICA—an identical spacecraft proposed by GE Astro Division.)

The Chinese have offered their recoverable FSW capsule to foreign users. This is an outgrowth—if not the identical twin—of their recoverable reconnaissance camera carrier. Thus it should be a reliable spacecraft. It has a payload of 300 kilograms and a staytime in orbit of about 10 days.

The Japanese have announced plans to fly their SFU space flyer in 1993 to carry experiments into space by rocket, with recovery by shuttle about 6 months later.

The Soviets have used their space stations Salyut and Mir as platforms for conducting various experiments in space. They have a recoverable capsule called Photon that offers 500 kilograms, 5 cubic meters, and about 3 weeks in space to foreign users. The French are under contract to use it. The American company Payload Systems is under contract to fly microgravity crystal experiments on the Soviet Mir. Their first flight took place in early 1990 and appeared to be successful. The company has the only license issued so far for flying American payloads on the Mir space station. The Soviets have offered a rather broad array of services to foreign users through their relatively new agency Glavkosmos—services including communications transponders, launch of payloads to geosynchronous orbit, microgravity experiments on Mir, and also photographs and radar images of earth made in space. I believe we should take advantage of the opportunity to use Soviet facilities more extensively—including even buying a Mir and running various simulated space missions on the ground. We could learn much from studying how the Soviets accomplish many objectives using less sophisticated technology than we normally employ.

All these foreign government facilities in operation or in the planning stages complicate decisions in this country about the commercial viability of proposals such as Spacehab. While foreign governments represent some of the best customers for such American initiatives, they also represent strong competition in the long run. They also raise once again the question of how private American companies can compete successfully against foreign governments. The answer must lie in offering a product or service that is somehow different from what those governments can offer. This reinforces the need for ingenuity, inventiveness, "creative financing," and other incentives but also indicates that head-to-head competition against such governments may not necessarily be a paying proposition.

■ SPACE HABITATS OF TOMORROW

Looking farther into the future, we need to plan for what are called closed-loop life-support systems. Rather than carrying along enough food, water, and oxygen to last for two or three years on a trip to Mars, we may be able to develop a space farm that will use human wastes to support crops of edible plants. Growing plants would not only convert waste materials into edible food but also would absorb carbon dioxide from the air and give off oxygen. Although we have taken a few plants into space to see how they react to this strange new environment, we have not run any extended experiments in space that would give us clues on how well such a thing might work. But we are now getting some valuable experience on the ground.

At a facility near Tucson, Space Biospheres, Inc., is building an experimental enclosure whose only connection with the outside world is electrical. A power line supplies electricity to operate test equipment and keep records of experimental results; it also is used for heating and cooling the facility. Aside from these connections, the habitat will be self-contained. In 1989, in a trial run, one person was sealed inside a 10,000 cubic foot facility for 2 weeks; the large facility—5 million cubic feet—is nearing completion and should be ready for operation by spring 1991. It will enclose 8 people for 2 years.

Experiments on life support systems are also under way at NASA's facilities at the Ames (California) and Kennedy (Florida) Space Centers. The Johnson Center in Houston is sponsoring such work at a number of contractor facilities. NASA's work has been of two types; one concerns the physics and chemistry of maintaining a good food and water supply and dealing with human waste products, including exhaled air, in space. Carbon dioxide is taken out of the air by passing it through lithium hydroxide cannisters.

Another group of scientists deals with biomass, trying to find ways to process human waste, grow plants, and reduce the weight of the food and other supplies that must be carried into space. It is not orbital flights that concern NASA most but flights beyond earth's orbit. A person needs about 14 pounds of supplies per day to keep going: 2 pounds of food, 2 pounds of oxygen, and 10 pounds of water. If a crew of 10 people went on a 3-year trip to Mars and back, they would need 150,000 pounds of stores, even if they never took a bath! And this assumes they would be willing to eat the last crumb of bread and breathe the last molecule of oxygen just as they reentered earth's atmosphere.

These large numbers make it desirable to find some way to operate on a closed-loop basis. The food becomes solid waste and urine, the water becomes urine and sweat and moist exhaled air, the oxygen becomes carbon dioxide. We know many ways to recycle all of these, but it is not done without cost—very high cost. Dr. Carolyn Huntoon at Johnson Space Center believes that we are unlikely to be able to develop a totally closed-loop system any time soon and must settle for a compromise where some recycling is done; other wastes are just stored away or discarded. As time goes on, we will find out how to process a larger fraction of the total waste products efficiently and thereby reduce weight.

Although the real problems are difficult enough, there is also a psychological problem. Waste water is certainly one of the obvious and easiest components to be recycled, but most people believe they would have to become rather thirsty to relish drinking reprocessed urine on a daily basis for 3 years.

Microgravity exacts its price on human beings. One of the most dangerous effects is bone demineralization, where a person loses about one half of one percent per month of calcium; there is also loss of muscle mass. The heart shrinks because there is less work to do, and the volume of blood decreases as the body's internal control mechanism misinterprets the extra blood it thinks is there. Production of red blood cells is cut back for similar reasons, and the red blood cells begin to take on strange shapes. When astronauts return to gravity, they find that they are very weak. It may take months to recover, and if bones demineralize too much, they may never recover.

All these things need to be studied and understood if we are to take long voyages in space. People arriving on Mars who have lost so much of their strength that they cannot deal with emergencies will not stand a very good chance of surviving. If they are to work on arrival, they must not be walking zombies, incapable of doing any heavy lifting.

Of course, one answer to this is the use of artificial gravity. But what fraction of the earth's normal gravitational effect is needed to maintain a vigorous body and mind? Studies on questions such as these need to be carried out when we have the proper habitat in which to do them. The space station is the obvious place to do the most extended studies, but the space station will not be ready until late in the nineties. Can we make do with other facilities in the meantime? The answer is a qualified yes.

We can do a lot of research in the shuttle using Spacelab and Spacehab. Several other additions to the shuttle have been specifically tailored to microgravity work. Two of these, to be mounted in the payload bay, are the Microgravity Space Laboratory (MSL) and the U.S. Microgravity Payload (USMP). The MSL covers about 50 square feet of deck space and can carry up to one ton of equipment. The USMP is twice as large.

■ THE PROSPECTS FOR COMMERCIAL SPACE HABITATS

The demand for habitats in space has thus far been limited primarily to research in the life sciences, but optimists say that research and development on materials processing will create a large market in space habitats in the future. Research and development, they believe, will turn up new materials that can be produced only in space, and will produce demands for places to carry out experiments on a much vaster scale than are presently under way. If there is enough work of this type there will be a demand for commercial facilities. The work on protein crystals (discussed in Chapter 8) is very exciting and may be the breakthrough we have been waiting for to create a larger demand for space workplaces and habitats.

Even if materials are not produced in space in large quantities in the foreseeable future, at least materials processing will be the basis for many experiments. And those experiments will in all likelihood keep the limited number of space habitats developed thus far tied up for years. There are some extremely interesting things to be learned both in life sciences and materials processing, and we must invest many months of work in space in understanding them. After that, many developments creating a need for new space habitats are likely. But until we can be sure, we should not oversell what we think we can do. Too much of that has happened already.

8

MATERIALS PROCESSING

A major activity of the potential customers for Spacelab, Space-hab, and other habitats in space will be research and development on materials and materials processes. In space we may be able to create new materials that are either impossible or uneconomical to produce on the ground. The better understanding of materials thus gained can also improve our ability to produce new materials on earth. Although much has been said about large-scale production of various materials in space, at this time there are no actual production programs waiting to be carried out in space. Nonetheless, many opportunities exist to do commercial experiments in materials processing in space.

One critical fact must be kept in mind, however: getting into space is expensive, so any product made in space must have a very high value per pound if it is going to become a commercial success. To take an extreme example, if a company discovered some process whereby sand could be turned into gold in orbit at absolutely no cost other than the cost of hauling the sand into space, the company still would not be able to make a profit, given the present cost of launches. Commercial success in materials processing will not occur until the expense of getting into space goes down, and many people are coming to recognize that more effort to bring this about is necessary.

Two properties of space account for the interest in materials processing in that environment: high vacuum and low gravity.

■ HIGH-VACUUM RESEARCH

Operating in a high vacuum means we can set up experiments where the presence of contaminants is minimized. Such a high vacuum may enhance the purity of alloys, drugs, and other materials. Although some experimenters have been interested in doing materials processing in space to take advantage of its high vacuum, it has not attracted a lot of attention because at the altitudes at which the shuttle and Space Station operate, the vacuum level is comparable to that which exists in a good scientific laboratory on the ground. Many would say that going into space for that level of vacuum is not worthwhile.

On the other hand, if one is doing experiments that generate much volatile material, or if one needs to maintain a high vacuum in a very large volume, being in space does have the advantage that the amount of material generated by the experiment is not likely to change the quality of the vacuum very much. It is the almost infinite virtual pumping rate of the vacuum environment that is useful.

To improve the "ambient vacuum" level in space at the shuttle's altitude, a new approach was put forward in 1988 by scientists at the University of Houston's Space Vacuum Epitaxy Center. The SVEC is one of 16 NASA-sponsored centers established at universities to work with industrial partners to engage in space-related research. The director, Dr. Alex Ignatiev, and his associates proposed using what is called a wake shield. An analogy to the wake shield is the cowcatcher on a locomotive that was used to knock cows off the tracks. The wake shield—now being developed by Space Industries, Inc., of Houston—was conceived to sweep out of the way particles that could contaminate a space experiment. If the wake shield performs as expected, the vacuum in space might greatly exceed in quality what we have generated on earth so far.

The Houston team needs a deep vacuum for their work on growing high-quality crystals. Such crystals can be useful in various electronic devices such as microcircuits used in very high speed computers. They are also useful to those trying to develop superconducting materials using molecular beam epitaxy techniques. Epitaxy is a process for growing crystals on a surface called a substrate, where the substrate itself is crystalline in form. Normally the material to be deposited on a substrate is heated in a vacuum to the point where it vaporizes. By properly designing the heating oven, one can cause the vapor to come off as a shaped beam. The substrate is then placed in the beam, where it intercepts the condensing vapor. The substrate itself must be heated to the right temperature so that the incoming vapor atoms can thermally accommodate themselves in optimal bonding configurations that result in crystalline thin film growth.

The whole process is a highly specialized art, and its success is strongly determined by the level of vacuum that is present. Any extraneous material may end up occluded in the surface of the substrate, whose properties are very sensitive to the chemical composition of the layer. In an ordinary lab operating under high-vacuum conditions, it is difficult to avoid having extraneous material introduced into the experiment. Going into space raises the possibility of achieving a much higher level of vacuum, with the actual quality of the vacuum related to how high one flies. At normal shuttle altitudes, the pressure is very similar to that in a good laboratory—a vacuum of 10^{-8} torr is considered good. A torr (from Torricelli) is a unit of pressure of $\frac{1}{1,000}$ of an atmosphere. If the shield works as planned, it could improve the vacuum significantly—to a level of 10^{-14} torr. Although the effectiveness of the wake shield has yet to be proven, it will, if successful, be a breakthrough in the level of vacuum achieved, and this in turn will enhance the purity of crystals grown there.

It is easy to imagine how the wake shield works in sweeping out of the way the particles resident in space. But what about contaminants being generated by the shuttle itself? One can envisage the shuttle as a habitat trapped in its own exhausts; anything thrown overboard just cruises alongside the shuttle. How can the wake shield deal with exhausts from the shuttle, and what about contaminants on the surface of the shuttle itself? Ignatiev and his associates say their method will solve all these problems. When the wake shield is in use, the orbiter will be oriented to fly "wing into ram," that is, sideways, with the shield deployed "upstream." Since the speed of the shuttle is 10 times the average thermal velocity of molecules (actually atoms, since molecules dissociate at these altitudes) encountered in space, it seems logical that there will be a shielding effect from particles outside the shuttle. There should be a cone-shaped volume of empty space behind the wake shield, since the shield is far enough above the mach line or shock wave generated by the orbiter that no orbiter contaminants can reach the protected volume. The shuttle will be flying sideways with the arm holding the shield above the shuttle and off to one side of the

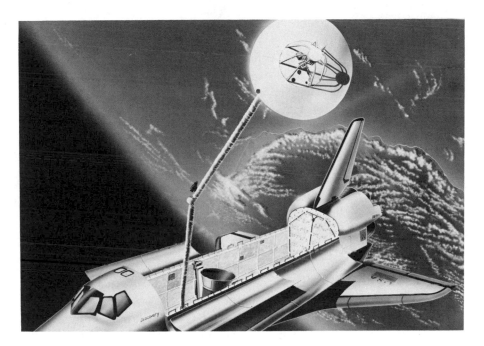

Figure 20. The wake shield, developed by Space Industries, Inc., will, if successful, make possible a much higher vacuum for scientific experiments conducted from the space shuttle. (Illustration courtesy of Space Industries, Inc.)

shuttle slip stream. With the shield flying ahead of the experiment package, there is almost no way for contaminating material to reach the experiment—unless it comes from the wake shield itself.

As for outgassing from the shield itself, it will be made of a molybdenum-rich stainless steel alloy, selected for its well-known favorable outgassing characteristics. Under normal conditions when the shield is not in use, it will be wrapped for protection from contaminants. All this is theory so far; when the actual flights take place, the orbiter and shield will be operated in various configurations to test the theory. If the design work progresses on schedule, the wake shield—together with the experiment package—will fly early in 1992.

The epitaxial thin films made possible by the wake shield should produce high-quality compound semiconductors that will lead to transistors 8 times faster than those used in current personal computers (PCs). The current PC market is approximately $4 billion. The new transistors could enable U.S. industry to capture a larger portion of the world market.

The wake shield facility is also expected to allow for the integration of semiconductor and high-temperature superconductor thin films for use in the manufacture of new logic devices. These could increase computer-processing speeds by a factor of 100 and increase the United States' lead in developing supercomputers.

Such techniques, if perfected, may not always require going into space, but the Houston team anticipates that an advanced electronic component manufacturing facility will be needed in space during the early twenty-first century. Accordingly, they have begun design work on such a facility. At the biennial Space Commerce convention in Montreux in March 1990, the team said they had done a study of a space-based facility to grow computer chips in production quantities, with such a facility being feasible by 1997—at a cost of $100 million. To justify such an expense, one would have to show that the resulting products would be worth the extra cost. Lest we assume that getting this amount of money is impossible, consider Motorola's microcircuit production facility in Austin, Texas; it cost $150 million. But we know that it will produce millions of units of such circuit boards—not the dozens Ignatiev will make in his space experiments.

While molecular beam epitaxy is an exciting field, we have not yet demonstrated that the shuttle can be a significantly better environment for its operation than a good laboratory. Much research and development must be done to see if this process can progress faster in space than on the ground—and at what cost. So it remains to be seen whether any such production facility can be justified in space.

▪ MICROGRAVITY RESEARCH

While the high vacuum of space is a good reason to want to do experiments there, an even more attractive feature of space is microgravity. Microgravity is a better term than weightlessness to describe the conditions that prevail in space. It is very unlikely that we will experience real weightlessness, given that the earth's gravity field varies with altitude, even over distances comparable to the size of the space laboratory itself. If there is a place in the laboratory that experiences zero gravity, other parts of the laboratory only a few feet away will not.

Typically, if there are no sources of vibration on board, a laboratory environment may vary between .000001 (1 micro) g and .000010 (10 micro) g. (One g is equivalent to the gravity one experiences on the surface of the earth.) Putting people and machinery on board introduces new factors. A space laboratory in a normal "quiescent" mode where people are trying not to create problems, and with machinery mounted so as to reduce vibration, may have an ambient environmental condition of .000001 g, but it is more likely to be in the range .000100 to .001000 g—a factor of 100 above that measured in the absence of these disturbers of the peace. Interestingly, many people planning experiments on the Space Station *Freedom* are assuming that they will experience microgravity levels of 1 to 10 micro g, but this may be more a hope than a reality.

In a microgravity environment many normally inhibiting effects are reduced. Metals, alloys, ceramics, and glass are all expected to yield new materials with new properties. When crystals are grown in a terrestrial laboratory, gravity can cause density changes that lead to convection currents. This effect is nearly absent in space. Consequently, alloys can be made much more homogeneous—the materials are not separating out because of density differences. We also expect that creation of entirely new alloys will become feasible. Moreover, structures are normally shaped by strength requirements imposed by gravity; take away gravity, and new structures of greatly extended dimensions may be possible. Limits on size of structures may be set by inertial considerations rather than weight.

In terrestrial laboratories we also normally experience hydrostatic pressure—the pressure encountered as one descends in a deep swimming pool—another missing element in space. Without hydrostatic pressure, other forces may become dominant, such as surface tension. A liquid contained by its own surface tension may be a good candidate for novel experiments. In space, containerless processing becomes possible, freeing the experiment from contamination and boundary effects.

Some of the most exciting experiments in space have been done by the pharmaceutical industry. Their work frequently involves arcane techniques where large quantities of materials must be processed to yield small quantities of drugs—some of which are extremely valuable. A few years ago, when people were first becoming excited about the possibility of producing new materials in space, James Beggs, as

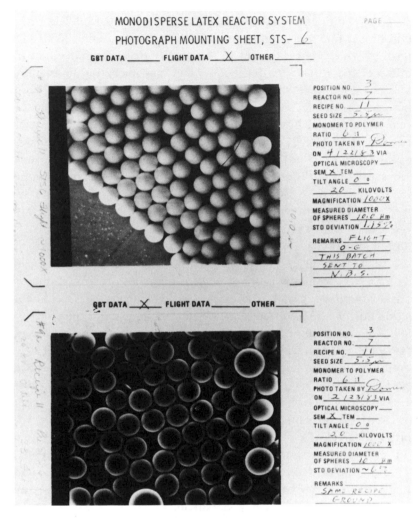

Figure 21. Some materials are more easily produced in space and to a higher degree of refinement, as seen in this comparison of electron microscopic views of monodisperse latex particles made on a space shuttle flight in April 1983. Space-produced 10-micron latex particles (top) are identical. Though made from the same chemical recipe using identical experimental hardware, particles made on earth under the influence of gravity (bottom) are deformed, off-sized, or imperfect in other ways. (Photograph courtesy of NASA.)

NASA administrator, gave many speeches about producing pharmaceuticals by a process called electrophoresis. It was said that a pound of such drugs produced in space would be worth millions. His optimism was based largely on experiments under way at McDonnell Douglas which were carried into orbit aboard the shuttle.

In 1978 McDonnell Douglas began looking for a way to become active in space commerce. Having been heavily involved in NASA space programs from the beginning—on Mercury, Gemini, and Apollo—they assumed that the best way to get into space commerce was to build something comparable to what they had built for NASA. This might take the form of a space habitat where people could do various kinds of microgravity work. After pursuing this idea for a time, they decided that there would be little future in it: once something had been built and launched into the relatively benign ambience of space, it would stay there forever. Better to go in for some kind of service that would have to be repeated frequently.

Jim Rose—now head of commercial activities at NASA and then at McDonnell Douglas—concluded that making drugs in space using electrophoresis was an appropriate entry in that market, since it would require repeated services as long as the drug was in demand. Since McDonnell Douglas had laboratories for life sciences and related fields, again the choice seemed to fit the capabilities already built up. Dr. Vernon Montgomery, head of the Life Sciences Division, was assigned to oversee the work.

Electrophoresis is a process that uses electric fields to concentrate charged particles and separate them by type or species within the carrier fluid—a critical process in the preparation of certain important new drugs. The McDonnell Douglas team selected a hormone called erythropoietin as the drug they would attempt to produce in space. Experimental runs of the equipment in space had shown promise of separating out the drug several hundred times faster than on the ground, and also with a higher degree of purity. If these results held up, going to space would make the process vastly more efficient and in turn might make a dramatic difference in the cost of the drug.

Rose and his team developed the equipment to separate out the drug within the carrier fluid, but key officials at McDonnell Douglas recognized that they knew nothing about marketing products of this type, and even though they felt confident about actually producing the drugs, they needed outside expertise about how to enter the pharmaceutical business. Success, they felt, would depend on tying themselves to a name-brand pharmaceutical partner. They asked Price Waterhouse to survey the field for partners they might work with. The Ortho Division of Johnson & Johnson stood out as a good prospect.

McDonnell Douglas submitted a proposal to NASA in November 1978 to do electrophoresis on the shuttle. Administrator Robert Frosch put an internal NASA task force to work on the idea in early 1979. That summer, the task force concluded that the process was a very good example of how the space environment could be used to do things which up to then had been impossible or very difficult. They also recommended that NASA absorb the marginal cost for carrying certain relatively small payloads into space via the shuttle. This cooperative approach was called a joint endeavor agreement (JEA), and McDonnell Douglas got the first such agreement in January 1980, about a year before the shuttle's first flight. They then made seven shuttle flights—four R&D flights in 1982–83, followed by three flights with a payload-specialist astronaut to oversee an experimental rig that was to engage in prototype production.

According to the agreement with Johnson & Johnson, McDonnell Douglas would make the product on the shuttle and deliver it to Johnson & Johnson, who would package and market it. The selected product—erythropoietin—is a hormone that stimulates the production of red blood cells, and a very important drug for those suffering from anemia, kidney failure, and other diseases. Johnson & Johnson identified erythropoietin as a drug with many applications, adding up to a vast potential—perhaps a billion-dollar market.

Test quantities of the drug were being made in the laboratory using natural cell cultures which produced the hormone at the rate of 200 biological units per liter. For an initial market size of 2.5 billion biological units, literally millions of liters of fluid containing unwanted material would have to be removed. Neither existing ground-based processing techniques nor electrophoresis operating on the ground could be scaled to meet such a demand. Initial tests showed that electrophoresis in space could process 500 times more material per hour than the technique produced on the ground; such an improvement might make the process practical and justify the high cost of going into space.

This looked like a breakthrough, and they went to work to carry out a production program. But they were working in a rapidly moving field; the product was important enough to attract other pharmaceuti-

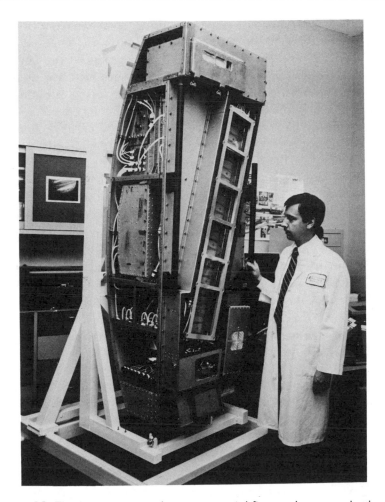

Figure 22. The first experiment by a commercial firm on the space shuttle was an engineering test of the Continuous Flow Electrophoresis System, designed by McDonnell Douglas in collaboration with the Ortho Pharmaceutical Division of Johnson & Johnson. The experiment was flown as part of a joint endeavor agreement (JEA) in which NASA and industry became partners in promoting development of commercial products in space. The companies agreed in advance to make the products derived from such experiments available to the public at reasonable cost. (Photograph courtesy of NASA.)

cal teams of experimenters. Soon a group of biotechnologists found a way to clone the gene of the hormone and grow the material in an established cell line on the ground. The cell produced the hormone with a density of a million units per liter. With such high-density material to start with, even the inefficient ground processes would be less expensive than more efficient processes that required going into space.

One company has recently succeeded in getting its biotechnology-based product approved by the Food and Drug Administration. Interestingly, it intends to offer the product at the same price that McDonnell Douglas predicted back in 1985. In effect the biotechnology group had scooped McDonnell Douglas's process and had shown how to bypass the expensive step of going into space to do this particular job.

Johnson & Johnson thought that McDonnell Douglas was on the right track 10 years ago, but J&J decided to drop out of the program when bioengineering companies entered the competition. Left without a partner, McDonnell Douglas began a search for a new one. Having confidence that the basic process was practical, McDonnell Douglas was ready to set up pilot production in space as soon as a partner was selected. Then came the *Challenger* accident and the whole program was put on hold. Many factors have kept that particular program on ice ever since.

But other medical programs are proceeding. Because of the importance of drugs to human and animal health, developing new drugs is one of the most active research areas in the world. Most drugs now available have been made by tedious trial-and-error methods. A more logical approach, which is the foundation of "rational drug design," is to learn the complete three-dimensional structure of a target protein using the techniques of crystallography, and then to specifically design compounds that have the proper shape and distribution of chemical groups to bind tightly within key sites of the protein—thereby modifying in a desirable way its biological action. These techniques may be referred to as drug design and protein engineering.

The study of protein crystals has expanded greatly as molecular biology has advanced; the molecular structure of proteins must be determined if we are to understand them well enough to learn how to modify their behavior. Since x-ray crystallography is almost the only way to find the 3-dimensional structure of protein molecules, the problem is to grow crystals which are large enough and pure enough that we can do x-ray crystallography of the necessary resolution. Many experiments have shown that we can grow much larger and more shapely crystals when gravity is absent.

Much of the advanced work on crystal structure is being conducted under NASA sponsorship at its various Centers for the Commercial Development of Space. At the Center for Macromolecular Crystallography at the University of Alabama at Birmingham, a team headed by Dr. Charles Bugg has now flown payloads 8 times in the shuttle's middeck to study crystal growth; they have been able to produce protein crystals in the microgravity of space that cannot be duplicated on the ground. The experiments carried out at the center in Alabama have used a technique called vapor diffusion. The protein is carried in a solution; the solution is pushed out of a syringe, forming a droplet on the tip of the syringe. The syringe is immersed in a chamber filled with a saturated solution of the precipitating agent; the protein solution is mixed with the precipitant by repeated extrusion of the droplet. The higher concentration of water in the protein- and precipitant-containing droplet causes water to diffuse through the droplet walls into the more saturated solution in the chamber, thus increasing the density of the liquid in the droplet. When a certain concentration is reached, protein crystals start to form in the droplet; in the absence of gravity the crystals formed are larger and purer than any that have been grown on the ground. They are brought back for laboratory analysis, including x-ray crystallography.

Work in space on several different proteins has been impressive and holds major promise for the pharmaceutical industry, even if researchers later discover how to produce these improved crystals in their terrestrial laboratories. The first protein is human gamma interferon, which stimulates the body's immune system and is used clinically in the treatment of cancer; its production has been investigated by Schering-Plough Research. Pancreatic elastase is a protein associated with the degeneration of tissue in people suffering from emphysema; Merck/Vertex is responsible for much of the commercial research on this chemical. Upjohn's work on pancreatic phospholipase holds promise for those affected by rheumatoid arthritis and septic shock. DuPont's work on isocitrate lyase may lead to the creation of fungicides for nematodes that cause serious crop diseases. Burroughs Welcome/BioCryst is working on reverse transcriptase, the protein that allows the AIDS virus to replicate itself, as well as other proteins.

Outside the pharmaceutical field, several other technologies that take advantage of microgravity are showing promise. NASA's centers at Batelle Columbus Laboratories and Clarkson University are working on improving catalysts used in industrial processes such as gasoline production. Some of the most important catalysts are zeolites—alumino-silicates. Knowledge and practical use of zeolites is not new; nonetheless, potential zeolite applications have been limited because of the difficulty of producing crystals of sufficient size and purity in the presence of gravity. Preliminary flight tests on the space shuttle have encouraged scientists to believe that zeolite production in microgravity will yield more perfect geometry and larger size zeolites. The first zeolites actually made in space were grown on a very short experiment performed on a sounding rocket flight—the Consort 3—in May 1990. That flight produced only 7 or 8 minutes of microgravity time. While the complete results are not yet available, preliminary indications are very promising.

The improved geometry of the zeolite catalysts will speed up many chemical processes that involve gas and liquid separations. Examples include polymer manufacturing, water purification, waste material management, drug production, and hydrocarbon processing (which could reduce the cost of automobile gasoline up to 30 percent). The zeolite industry grosses almost $1 billion per year, of which about 70 percent is in ion exchange and separation processes. Improvement of zeolite crystal production could cut these processing costs in half—which would double or triple the business.

Another example of space-oriented materials research is the Minnesota Mining and Manufacturing Company's work on the Physical Vapor Transport of Organic Solids (PVTOS). This privately developed experiment to produce organic thin films with ordered crystalline structures has led to a patent application by 3M. Thin films have optical, electrical, and chemical properties with significant potential for commercial applications. All nine samples that 3M has flown on the shuttle have produced films that were smoother and more highly ordered than ground-produced films, and thus have more homogeneous optical properties.

The 3M polymer morphology experiment, flown on the shuttle in October 1989, was the first of 62 proposed flights for 3M under a 10-year joint endeavor agreement with NASA. The work is part of a sophisticated investigation by 3M into the effects of microgravity on the processing of polymeric materials. Melt processing in microgravity eliminates the influences that sedimentation and buoyancy-driven convection have on the physical properties of polymers. This commercially designed and constructed experiment equipment uses infrared spectroscopy to study the structures of the organic materials as they undergo a transition or change in state. Although data analysis was not complete at this writing, all indications are that the equipment operated properly, with 93 percent of the data being collected.

Another recent development is the Fluid Experiment Apparatus (FEA), sponsored by Rockwell International, which first flew in 1984 as part of the Shuttle Student Involvement Program. It flew again in April 1989 on the first of two flights in Rockwell's joint endeavor agreement with NASA. The experiment used indium samples to study floating zone crystal growth and purification. The second flight, which took place in January 1990, studied the effects of microgravity disturbances on floating zone crystal growth—a method also used on the ground. This method of processing involves melting a portion of a rod-like sample and moving the molten portion along the length of the sample. As the sample material cools and resolidifies behind the heating unit, it forms a single, pure crystal. The FEA also included a Microgravity Disturbance Experiment (MDE) designed to study how orbiter and crew-induced disturbances affect stability in the molten zone used in float-zone crystal growth.

The FEA payload carried samples of commercial-purity indium provided by the Indium Corporation of America, which is collaborating with NASA in developing and analyzing the experiments. Indium is a well-understood material, with a relatively low melting point, used in the production of electronic devices. Initial results from the FEA payload are very positive, with all seven main indium samples being processed on the flight, as well as part of the spare eighth sample.

All crystal growth experiments conducted in the United States up to now have flown on vehicles with very limited duration in space. The shortness of exposure time has been made up for by seeding crystals on some shuttle flights. But of course the seeding technique can be applied to flights of any duration. Once the Space Station *Freedom* is in orbit, researchers will have years instead of a few days to try out various crystal-growing techniques.

While we eagerly await longer flights in microgravity, the Soviets have been able to conduct such flights since the Mir spacecraft was first launched in 1986. While undoubtedly much good work has been done on the Mir, our knowledge of it is not extensive. Recently, Dr. Byron Lichtenberg of Payload Systems in Cambridge, Massachusetts, contracted to fly 6 separate payloads on Mir, the first of which returned from orbit in early 1990. Payload Systems has completed 50 projects related to microgravity in three different environments—the KC-135 aircraft, an Air Force tanker operated by NASA, which flies a hyperbolic trajectory, achieving about 30 seconds of reduced gravity time; the space shuttle, which typically stays in orbit for a week; and the Mir spacecraft, where a 56-day experiment was run.

In this experiment, a number of vapor diffusion samples were exposed to a microgravity environment for a much longer time than in any previous U.S. experiment. As of this writing, no quantitative information is available on the Mir flight, but experimenters are generally optimistic.

■ RADIATION

While working in the vacuum and microgravity of space may be advantageous to some experimenters, another property of space that comes with being above the atmosphere may be highly detrimental. When we journey above the atmosphere, we lose the protection from radiation that the atmosphere normally provides. This property of space does not usually attract much attention, but it is very important to keep it in mind if people are to remain in space for very long periods of time.

Streams of charged particles emitted by the sun in what is called the solar wind, as well as cosmic rays—high-energy particles arriving from outer space—can be very harmful to living matter and other sensitive materials. The effects of radiation in space are normally ameliorated by use of proper shielding materials that absorb it, by avoiding certain altitudes where radiation is most intense (the Van Allen Belts), and by creating special compartments (like storm cellars) into which crews may go in times of solar flares or other unusually high radiation levels.

Thus far, radiation has not been a major problem for crews on the shuttle, whose maximum stay in orbit has not exceeded 10 days. Skylab crews did not suffer ill effects from radiation even on the longest voyage of 84 days. And cosmonauts who have been in space for a full year have not shown evidence of damage from radiation. The radiation dosage received by astronauts (and cosmonauts) is routinely monitored; if they were to accumulate very strong radiation dosage by flying during periods of excess solar activity, their future missions in space would be limited so that their lifetime exposure was held within tolerable levels. But precautions must be taken when one is in space because there is little warning of intense solar activity. There is considerable doubt whether we know just what levels of radiation might be received by a person on a year's voyage to Mars, for example.

Unfortunately, NASA has been less than successful in persuading Congress that much more work should be done on this subject before we embark on a voyage long enough that a quick return to earth is impossible. Some critics of manned space flight on Capitol Hill are said to be using their influence to hold down the amount of research on this subject; this could be an indirect way of slowing the pace of flights to the moon and Mars.

For research purposes, this "ambient" radiation may have a positive side, providing opportunities for studying the effects of atomic oxygen that is present in this region of space. Normally oxygen does not exist in the atmosphere in atomic form; two atoms of oxygen combine to form a stable molecule. But high-intensity radiation from the sun (such as ultraviolet rays, which are normally stopped by the atmosphere) causes dissociation of the two atoms. Consequently, the top of the atmosphere contains only atomic oxygen—a form of oxygen which is very active, even caustic. The space environment will allow us to conduct experiments in high-energy oxidation chemistry. Such work is now under way at NASA centers at the University of Houston, the University of Alabama at Huntsville, and at Case Western Reserve University.

▪ FUTURE OUTLOOK

The "obvious" example of materials processing in space that NASA backed for years—electrophoresis—did not pan out. Now the great interest is in proteins, with the object of growing very pure crystals that will reveal their basic molecular structure. This knowledge will, in turn, enable the creation of new molecules with interesting and perhaps beneficial properties. Whether we experience a new disappointment with respect to proteins or not, an excitement about materials processing has endured through the hard times of no access to space and should continue to grow as opportunities expand. NASA's Office of Commercial Programs, and its Centers for the Commercial Development of Space have rejuvenated activity in this field, involving dozens of universities and more than a hundred U.S. firms. U.S. researchers are seeking useful applications in many technologies making use of the limitless vacuum, low gravity, and radiation found in space. Efforts to increase access to space and reduce its costs will add to the likelihood that profitable uses can be made of what we are learning.

There probably will not be an early need for large-scale materials processing for production purposes, but just the experimental work that can now be done will keep scientists and members of R&D departments busy for the next 5 years. Such work is a justifiable end in itself and will keep a number of space facilities occupied. It is the only way we can answer the question of whether going into space can be commercially viable. Should we find that we can now do experiments in space that are markedly superior to what we can do on the ground, this would be a very significant development both for science and for space commerce.

9

FINANCING PROJECTS IN SPACE

Space commerce cannot proceed—much less succeed—unless there is money available to finance it. Except for the handful of cases where a wealthy person wants to make a name for himself in space, the usual mix of equity and debt capital must be brought together to make space enterprises viable. It is fair to ask just what makes space commerce more attractive or less attractive than the run-of-the-mill business activity.

In some respects space ventures and aviation are quite similar kinds of businesses. In watching the aviation and airline industry for about 45 years, I have noticed that there are reasonable numbers of people who think aviation is a good business to be in simply because they are attracted to the miracle of flight. Overcoming gravity and flying up there with the birds has a special romantic appeal, and many people are willing to invest their money and effort trying to make aviation pay off. This has been good for aviation, because many times in the past, decades have gone by with few if any signs of financial payoff in aviation. Yet those who stayed with it usually made out very well in the long run. I think the same is true about people's attraction to investments in space; a certain romance is connected with spaceflight, because traveling into space defies gravity on an even grander scale.

On the other side of the ledger, there is a major factor which makes space commerce less attractive than many new business ventures on the ground, and that factor is political risk. As a Department of Commerce booklet, *Commercial Space Ventures* (April 1990), points out, two major types of risk in space are technical risks and market risks, both of which are very familiar because they apply to most businesses.

Political risks, however, are much greater in space than on the ground and are less familiar to most businessmen. They may well be the single most significant barrier to financial investment in commercial space ventures.

Political risks arise because most commercial space ventures include the government as a key player—either as a partner, as an important customer, as a supplier of a critical need, or as a regulator of activities. While the private parties involved can deal with the normal risks—and the financial community can decide whether what they propose to do constitutes a reasonable business opportunity—the government may suddenly change its mind about policy or priorities and leave a fledgling business venture stranded. A case in point, as we have seen, is Orbital Science Corporation, which placed its original bets on TOS (an upper stage for the shuttle), only to have the shuttle go out of the commercial launch business after the *Challenger* loss. Changing political ideologies about the proper mix of private and public ownership of national assets, particularly those that are related in some way to national security, can also have dramatic and catastrophic effects on space industries.

Spacehab represents another kind of political risk in that this module has only one purpose and that is to increase the habitable space in the shuttle. If NASA wants to use it, there is a role for Spacehab; if NASA does not care for the concept—or at some point ceases to care for the concept—Spacehab is dead. Spacehab literally cannot get off the ground on its own, and since it must fly on the shuttle, the government (read NASA) can set the price of transportation at whatever level it wants to, depending on whether it sees merit in having Spacehab succeed as a commercial venture.

Spacehab, from its earliest days, has been dependent on the government for its viability. It was originally conceived as a passenger carrier to put people in the cargo bay of the space shuttle. When NASA said no to that idea back in 1984, Spacehab became a laboratory where microgravity work could be done. Spacehab's founder, Robert Citron, raised a few hundred thousand dollars from friends to carry out feasibility studies and then raised about a half million to stay alive another year. Then the new president, Richard Jacobson, helped raise $1 million from friends and space buffs. Spacehab next formed relationships with Aeritalia (the company which had built Spacelab for ESA) and with McDonnell Douglas. These aerospace companies have done much of the design work on the module without demanding a lot of cash. In late 1987 Spacehab managed to raise about $2 million through a venture capital firm to continue its design work. And now with James Beggs as chairman, the company has raised about $140 million—$29 million in equity and the rest in debt led by the Chemical Bank of New York.

It appears that Spacehab's financial needs will be met and that Spacehab will soon have its chance to become a successful space venture. Many of us thought that was the case a year ago, however; who knows if there are other hurdles out there awaiting the unsuspecting partners of Spacehab. As of September 1990, Spacehab was awaiting a certification from NASA that it would indeed be allowed to fly on the shuttle in 1992–93. Of course flying is not the only thing it must do; it must get enough customers who want to use its services to make the whole venture pay off. Whether that will happen remains to be seen.

▪ FINANCING COMMUNICATIONS SATELLITES

The most frequently financed space projects have been communications satellites, since they are the bulk of space activity. Comsat benefited by being the first company in the space business; a high profile helped it raise its first capital through a $200 million stock issue—an issue that was immediately oversubscribed. When Comsat wanted to set up Comsat General to operate a domsat business a few years later, it had $200 million in the bank with which to do the job. But when Comsat later tried to raise money for a direct broadcast system, it ran into the kinds of problems most companies face when they ask investors and banks for money: will DBS make a profit? Comsat's inability to persuade enough of the right people that it would do so had a lot to do with the abandonment of DBS by Comsat. Comsat decided rightly that it should not invest $500 million in such a business on its own because that would be like betting the company on the endeavor.

The next U.S. companies to go into space—RCA and Western Union—were established businesses with good credit lines; they had little trouble finding the money. Typically companies have needed to raise $100–200 million to go into the communications satellite business—an amount which large companies can take in stride and which small companies have to work very hard to raise. GE, AT&T, GTE,

and Hughes—the large companies who have succeeded in space communications—have had no special financial problems. Smaller companies like American Satellite, Direct Broadcast Satellite Corporation,
and Orion Satellite Corporation did not have the assets of the large
corporations and were more in need of creative financing.

In addition to those corporations that have actually put up satellites,
there have been dozens of companies which thought they were going
into business but never did. Many of them—especially the plethora of
companies that wanted to go into the domsat business—found that
they were not attractive to potential investors.

How does a potential investor look at a communications satellite in
order to assess its likelihood of making money? According to Jerome
Simonoff of Citicorp, who has had a great deal of experience related to
financing space ventures, from the perspective of a potential investor a
satellite has to be able to recover four major costs if it is to be profitable: (1) the cost of the spacecraft, (2) the cost of the launch, (3) the
cost of insurance, and (4) the cost of money over the life of the project.

To assess these costs and the likelihood that a given satellite will
recover them and go on to make a profit, Simonoff says it is helpful
to think of a satellite as a piece of real estate. For example, many
satellites sit still in the sky, occupying slots assigned to them by federal
authorities. Since satellites generally have 12–30 transponders on
board, transponders can sometimes be sold off one at a time (like
condos) to raise money to build the satellite. The value of a transponder is determined by its location and capacity, much as a piece of real
estate is evaluated according to its location and square footage. In the
case of a satellite, the question is how many people—and specifically
which people—on earth can see it. Some satellites live in more affluent
neighborhoods than others, and that increases their value. When there
happens to be a shortage of satellite capacity in orbit, as there was
about 10 years ago when everyone suddenly wanted to feed more
signals to cable TV systems, then the value of a transponder in the
right place at the right time might exceed what it cost to put it there
by a very large factor.

Simonoff points out that only equity investors are willing to gamble
on the commercial feasibility and technical quality of space systems;
lenders do not come in until they see investors putting up enough
equity capital to show their commitment to the project. Management
cannot borrow all the money it needs from a bank; banks want a high

degree of certainty that they will be paid back and hence want to see that investors are betting substantial amounts of money on the project's viability. Banks also want to know what the venture's hard assets are, in the event the business does not succeed and the assets have to be sold. Space hardware may not be regarded by banks as readily saleable; ground-segment hardware may or may not be more saleable.

Normally, both investors and lenders want to see the project adequately insured. But lately some companies have been tempted not to buy insurance because the large numbers of launch failures in the last few years have sent insurance rates sky high. Recently Intelsat took this gamble on an Intelsat 6 scheduled for launch in early 1990 on a Titan 3 rocket. When the launch failed, the net result was that Intelsat and Martin Marietta ended up suing each other.

Because of the time-value of money, investors want to know when the system can be expected to begin earning money. Typically, space hardware takes 3–5 years for construction and launch; a company's ability to shorten the time needed may make a large difference in the attractiveness of the project to investors, and a subcontractor's ability to deliver on a tight schedule may increase the attractiveness of a supplier's bid.

▪ TWO CASE HISTORIES: GEOSTAR AND QUALCOMM

The cases discussed thus far have been conventional communications satellites, which carry telephone traffic, television, data, and the like. Two companies that are trying to make a profit by using communications satellites for a new purpose are Geostar and Qualcomm. Both of these companies propose to use satellites for managing truck-tracking systems.

Qualcomm (described in Chapter 6) was founded by veterans of the communications business who wanted to do R&D on new and interesting concepts under government contracts. After succeeding in this business, they decided to put together the OmniTracs truck-tracking system. Since they did not need dedicated satellites but could lease existing Ku-band capacity, their financial needs were not great. The founders raised some money from fellow workers and colleagues and reportedly also used the investment bankers Goldman Sachs.

Geostar, by contrast, required dedicated satellites or at least dedicated packages on satellites; they could not just take whatever satellite capacity happened to be available. This meant raising major amounts of money. To get started, the system's inventor, Gerry O'Neill, contacted friends and raised some $2 million in private financing in 1983. This was used for R&D to prove the validity of the concept. The next $2.5 million was raised by Wheat First Securities of Richmond, Virginia, in a private placement in 1984.

Having estimated that some $300 million would be needed to put the basic Geostar system in place, Geostar executives were told by First Boston that it was unlikely they could raise that amount of money from scratch for the proposed purpose and that they should find a way to go into business on a phased approach. Geostar came up with a plan which permitted limited operations but demonstrated that their system of truck tracking and display would work; this would generate a certain level of revenue. They would then be in a better position to ask for the full amount and as a consequence could grow step by step to the originally planned configuration of 3 satellites and a 2-way communications system.

The basic equipment needed to demonstrate the system was financed and built—with some $75 million being raised in the process. Unfortunately, Geostar was immediately plagued with technical problems. The first satellite package they put in orbit lasted only 6 weeks, the second package was a year late getting into orbit, and the third went up on a satellite that was placed in the wrong orbit, necessitating much extra work and time and loss of control gas before it was operational. Because of all these events, financing for the full system has not yet been raised.

One very positive development for Geostar is an arrangement with CNES, the French space agency, whereby CNES orchestrated a 10-nation consortium to create a company called Locstar to offer services similar to Geostar's in Europe. If Geostar can achieve certain milestones in providing service and earning revenue, the CNES-generated consortium is a good possibility for helping to finance Geostar.

Geostar's business is in a field of space commerce which in some ways should be easy to finance, since it is a variation on the old theme of communications satellites. Geostar is basically a communications system specializing in locating moving vehicles, then reporting their positions and exchanging messages between trucks and truck-fleet operators. But Geostar is unconventional enough to raise the question whether experience on the other satellite systems can be extrapolated to this case. The jury is still out on whether Geostar can reach the point of financial viability.

■ FINANCING THE LAUNCH BUSINESS

Conventional communications satellites have come to be taken for granted as successful space ventures—so much so that many people do not even mention communications when they speak of space commerce; they have in mind the more recent activities such as remote sensing, navigation, habitations, and materials processing. These newer activities do not have the solid historical foundation that communications satellites enjoy, and financing them is often a much more difficult undertaking. The commercial launch business, on the other hand, has much in common with the satellite communications business. As a profit-making enterprise, it is second only to communications in revenue; and the long history of successes that the major builders of rockets enjoyed while working primarily for the government established them as good risks when they went into the commercial launch business.

In the case of the large launch companies, financing is not a primary concern. General Dynamics, McDonnell Douglas, and Martin Marietta are big enough that getting money to finance their commercial launch business has not been a problem for them; their main question has been whether the U.S. government will control the foreign competition, so that U.S. companies can make money (see below).

The financial situation of several small launch companies was touched on in Chapter 4. Orbital Sciences Corporation's seed money came from a few interested private investors in 1982; then several venture capital firms invested $2 million in 1983, after which Shearson Lehman Hutton, an investment bank, was able to put together an R&D partnership for the bulk of the money needed for OSC's first project. OSC's money managers were able to find 1,000 people willing to invest $50,000 each—a rather powerful way to raise money. Tax policies of the time (1983–84) made financing the venture much more attractive than it would be today because of recent tax law changes. For their second rocket development, OSC sold 20 percent of the company to Hercules—the principal supplier of its rocket motors—which raised enough money to develop the Pegasus rocket launcher. Finally, in 1990, OSC "went public," selling $36 million of its stock to investors.

Space Services, Inc., another small launch company, planned to offer launch services similar to OSC's. Unfortunately, its money came principally from one private source, and when that source dried up, it was left without capital. Other small launch companies have had similar problems because they have had to depend largely on private investors with a particular interest in being in the space business. When the interest of these investors wanes, the small companies often find themselves out on a limb.

For such companies with little money of their own, a source of government funding is very important. Enter DARPA, the Defense Department agency whose role is to see that new technologies are made available for defense needs. Many such needs span not only defense but also civil space, and so DARPA and NASA frequently are looking at the same technological development possibilities. DARPA, as we have seen, has run several competitions for launchers of payloads under 1,000 pounds. NASA is also showing interest in companies that can launch small satellites, which are currently enjoying a rebirth of interest. For the several small companies wanting to play in this field, winning one or more such launch opportunities has become very important to staying alive.

▪ PUBLIC/PRIVATE PARTNERSHIPS IN SPACE

Space commerce, as we have seen, faces a set of problems that is unique in the business world. It must deal with international competitors who do not play by the same ground rules for raising money that we in the United States have traditionally espoused. It is plagued with political uncertainties that arise in large part from the importance of space activities to our nation's military interests. And compounding those political risk factors are the extremely high costs and technological risks of operating a business in space. How are these obstacles to be overcome?

James Harlan Cleveland, former ambassador to NATO, has written extensively on our society's need to develop new mechanisms to carry out large projects which serve a grand social purpose. He sees new problems arising which our past practices—let alone our founding fathers—did not contemplate. To deal with such problems, he has long advocated a larger role for what he calls public/private partnerships. In *The Global Commons: Policy for the Planet,* he deals with issues like how we should think about the assets we all hold in common and how we should expand them as we try to bring the benefits of modern technology to bear on the common problems of mankind.

Many European countries today seem to be moving toward something resembling Ambassador Cleveland's definition of public/private partnerships in their telecommunications industries. In the past, government domination of these industries was the norm. When Intelsat—the first international satellite communications system—was being established, Comsat was one of a very few private companies acting as their country's representative. Typically the investors were the telephone companies or PTTs of the member countries, essentially all of which were divisions of government ministries. Recently, however, many nations have begun the process of liberalizing the government's control of telecommunications, and many countries are freeing up their PTTs to enable them to be more aggressive and operate more like private businesses.

British Telecom (BT) is a recent example. It has gone from being totally owned by the government to being 51 percent owned by the public, with the government's shares not normally voted. Thus BT's control is in the hands of the private sector. BT was noted for its lethargic approach to putting a modern communications network in place. But things are now very different. In addition to putting BT under private control and ownership, the British government created Mercury—a small but dynamic competitor for BT. In effect, they created a duopoly in an attempt to have the best of two worlds. They wanted competition; hence two companies were needed. They also recognized that in many areas service would have to be provided at no profit; hence the need to maintain the concept of the regulated utility.

A BT subsidiary, British Telecom International, or BTI, is the British participant in Intelsat. It too has gone from state ownership to private ownership. Even in France, where the term PTT originated, we see movement in the direction of privatization. The old PTT is now the PTE for Post, Telecommunications & Space (Espace). Its major arms are DRG (Directorate of Regulatory Affairs) and France Telecom, which is the operating arm. As time goes on, France Telecom, like many other PTTs, is expected to operate more and more like an autonomous private organization.

More than a dozen countries are in the act of privatizing or at least freeing up their PTTs. For example, Teleglobe of Canada, the country's representative in Intelsat, was sold to the private company, Memotec, Inc. Jamintel in Jamaica has been privatized, with 67 percent owned by Cable & Wireless and the remaining 33 percent belonging to the government. Radio Marconi in Portugal operates the international service under license from the government. And in the Dominican Republic, GTE owns the telephone company and operates under government license. DBP Telekom in Germany is a government-owned corporation, as is PTT Telecom in the Netherlands. Telespazio in Italy is a private company owned by STET, which is largely owned by the government.

While many foreign governments are allowing their PTTs to become more like private entities and are privatizing their international arms by selling them to the public or to private companies, these governments are also allowing non-Intelsat affiliates such as PanAmSat to provide service within their borders and compete with government-owned industries. The British government, for example, is allowing British Satellite Broadcasting to offer DBS services.

All these examples illustrate the point that when it comes to space commerce—including but not limited to satellite communications and launches—governments are either the source of business or the arbiter of the competition. The decisions of governments are crucial to success in space in any country, including the United States. Yet governments around the world, from Mr. Gorbachev's to Mr. Mitterand's, are recognizing the need to use the private sector and its methods of operation to stimulate their economies and get the job done. The recent changes in the PTTs illustrate a widely held belief nowadays that governments do a poor job of operating businesses.

Aerospace industries in foreign countries have moved more slowly toward privatization than the telecommunications industries. In Britain, for example, governments have moved back and forth on the issue of whether the aerospace industry and the airlines should be in public or private hands. These industries have been nationalized, then sold back to the public. In France Airbus Industrie and Ariane are still prime examples of industries owned and controlled by government. This mode of government ownership does not match what we are accustomed to in the United States and yet we are forever confronted with dealing with these entities on a competitive basis. Can corporations, investors, and the government in the United States achieve some sort of new modus operandi which will permit us to compete successfully with foreign nations in this changing environment?

I believe the answer is yes, provided we are willing to make a few adjustments in our attitude toward private enterprise. First, we need to reach formal agreements on how we share the market with our government-financed competitors. Discussions on this subject have already been begun with ESA. The very idea is anathema to many people, but there may be no other choice.

Second, we must be willing to engage in public/private partnership arrangements unlike anything we have known in the past. For the large companies to stay competitive, as we have seen, what is needed is a steady stream of orders for hardware and services of various kinds. This stream of orders has been adequate to date in the case of shuttles, satellites, ground stations, and launches of government payloads. But in the field of remote sensing, for example—with the exception of the transitional Landsat—the government is still buying satellites and collecting its own data. I believe it is time for the government to consider public/private partnerships which would permit the government to buy data from the private sector; if NOAA bought weather data from the company operating Landsat, for example, Landsat's income would be large enough to remove all doubt of its success. Similar opportunities to buy data from the private sector could be created under the auspices of EOS and Mission to Planet Earth. Both of these activities also lend themselves to international operation and support in a public/private organization similar to Intelsat. I hope to see some such arrangement materialize in the next few years.

What the smaller and startup companies frequently need is a chance to show what they can do. This applies to companies developing habitations in space, navigation systems, small launchers, materials processing experiments, and life science studies. Orders for their wares from DARPA and NASA have already provided much-needed opportunities, but much more could and must be done. In many industries outside of communications and the launch business, it appears that there will be no significant space commerce in the United States unless government policy rises to the occasion to provide that essential underpinning of orders. Perhaps when these other fields of endeavor have progressed and matured sufficiently, they will no longer be dependent on the government as a customer. But for now, in spite of my usual optimism, I must say that I do not see any ready examples of stand-alone business successes in space. All must count on government orders to supplement their commercial business.

SOURCE NOTES

The works cited here are only the main sources used in the preparation of this book; these notes are therefore not a definitive guide to the references used. Many of these books contain extensive bibliographies.

Periodicals which may be useful to those interested in space activities include the following: *Ad Astra* (monthly; Washington, DC: The National Space Society); *Aerospace America* (monthly; Washington, DC: The American Institute of Aeronautics & Astronautics); *Air & Space* (bimonthly; Washington, DC: The Smithsonian Institution); *The Annual Satellite Directory* (Potomac, MD: Phillips Publishing, Inc.); *Aviation Week and Space Technology* (weekly; New York: McGraw-Hill); *Satellite Communications* (monthly; Englewood, CO: Cargill Publishing Co.); *Space* (bimonthly; Burnham, Buckinghamshire: The Shephard Press, Ltd.); *Space News* (weekly; Springfield, VA: The Times Journal Company); *Space Policy* (quarterly; London: Butterworth); *Space Times* (bimonthly; Washington, DC: The American Astronautics Society); *Spaceflight* (monthly; London: The British Interplanetary Society); *Spacewatch* (monthly; Colorado Springs: The United States Space Foundation); *Via Satellite* (monthly; Potomac, MD: Phillips Publishing, Inc.).

Introduction

By way of background to recent times, Wernher von Braun and Frederick I. Ordway's well-illustrated *History of Rocketry and Space Travel,* 3rd rev. ed. (New York: Crowell, 1975), covers all the actions and systems mentioned in this chapter, including Project RAND, the ballistic-missile race with the USSR and its consequences in the early years of the space program, and the buildup to the Apollo program and its lunar operations. See also Frank H. Winter, *Rockets into Space* (Cambridge, MA: Harvard University Press, 1990). William S. Bainbridge's *The Spaceflight Revolution* (New York: Wiley, 1976) registers the optimism that inspired many new companies and the euphoric "Paine Report," *Pioneering the Space Frontier,* issued by President Reagan's

National Space Commission, chaired by ex-NASA Administrator Thomas O. Paine (New York: Bantam, 1986). But see also the *Report of the Presidential Commission on the Space Shuttle Challenger Accident* (Washington, D.C., 1986) and "Fact Sheet: The President's Space Policy and Commercial Space Initiative to Begin the Next Century" (White House Office of the Press Secretary).

The Space Business Research Center in Houston, Texas, compiles figures on military and civil space industry and publishes them together with market assessments in its *Space Business* report, which covers U.S. government programs, space transportation, communications satellites, remote sensing, and microgravity processing. In "Making the Grade" (*Omni,* April 1988, p. 71), Jerry Grey of the American Institute of Aeronautics & Astronautics (AIAA) assesses the rising expectations and competence of other nations in space research and vehicle deployment. The AIAA Position Paper, "The U.S. Aerospace Industry and America's Competitiveness" (Washington, DC: AIAA, September 1989), portrays the relative R&D investments of France, Japan, West Germany, and the United States. M. Harr and R. Kohli, *Commercial Utilization of Space—An International Comparison of Framework Conditions* (Columbus, OH: Battelle, 1990), describes existing programs, economics, and organizational, legal, and political conditions affecting competition, and includes a substantial bibliography.

1. The Early Years of Space-Based Communication

Arthur C. Clarke's imaginative proposal for geosynchronous positioning of relay communication satellites appeared in "Extraterrestrial Relays," *Wireless World,* October 1945. It attracted the attention of cognoscenti only but may have influenced the contributors to Project RAND as they mapped satellite applications in 1946; see "The 1947 Lipp Report on Satellites for Ocean Surveillance, Reconnaissance, and Geostationary Communications," in Merton E. Davies and William R. Harris, *RAND Corporation's Role in the Evolution of Balloon and Satellite Observation Systems and Related U.S. Space Technology* (Santa Monica: RAND, 1988)—an important source book. Nevertheless, ten years later a conventional scientist, John R. Pierce, Director of Electronics Research of Bell Telephone Laboratories, could still describe conditions and tradeoffs for "Orbital Radio Relays" (*Jet Propulsion,* April 1955). After the launching in July 1963 of the experimental geosynchronous satellite Syncom 2, a special issue of *Astronautics and Aerospace Engineering* (September 1963)—except for the dissenting views of Pierce—emphasized the advantages of equatorial synchronous satellites.

In the February 1964 issue of *Astronautics & Aeronautics,* Arthur Clarke, writing from a fishing village in Ceylon (now Sri Lanka), could confidently predict not only telephone and TV transmissions through geosynchronous communication satellites but also an orbital post office, orbital newspapers, and an electronic blackboard. To the informed lay public, Burton I. Edelson's "Global Satellite Communications" in *Scientific American,* February 1977, described the ascendancy of geosynchronous satellite communications (followed soon by Pierce's change of opinion). Clarke's *Voices from the Sky: Previews of the Coming Space Age* (New York: Harper & Row, 1965) clinched the picture. Delbert D. Smith's *Communication via Satellite: A Vision in Retrospect* (New York: A. W. Sijthoff, 1976) added the wisdom of two decades' hindsight. Edelson's "Communications Satellites—The Experimental Years," *Acta Astronautica,* vol. 2, nos. 7–8, pp. 407–413, concisely summarizes the contenders of the early sixties, as does the succinct opening chapter of Wilbur L. Pritchard and Joseph A. Sculli, *Satellite Communications Systems Engineering* (Englewood Cliffs, NJ: Prentice Hall, 1986). The early history of space-based communications, the technical politics, the "experimental years," the role of Comsat, the Satellite Act, and the movement to the modern global geosynchronous systems have been conveniently and authoritatively summarized, together with systems characteristics and the outlook for further developments, in *The Intelsat Global Satellite System,* ed. Joel Alper and Joseph N. Pelton (New York: American Institute of Aeronautics & Astronautics, 1984).

2. International Communications Satellites

President John F. Kennedy called for the development of an international communications system in space in his May 25, 1961, speech before a special session of Congress on "Urgent National Needs." Subsequently, the legislation creating Comsat and the international system was the focus of an intense legislative debate over the mission and structure of such a new business-government partnership. While the list of hearings is too extensive to cite here, in the year 1962 the Senate Committees on Commerce and Foreign Relations and the Senate Select Committee on Small Business, as well as the House Committees on Interstate and Foreign Commerce, Judiciary, Science and Astronautics, Foreign Affairs, Government Operations, and Science and Technology, all convened hearings on various aspects of the 1962 Communications Satellite Act. Actual debate on the Act can be found in the *Congressional Record* dated from April 25 to August 30, 1962. The Communications Satellite Act was signed into law on August 31, 1962.

There have been several popular critiques written about the Comsat Act and its impact on satellite communications development and technology. Among these are: Michael Kinsley, *Outer Space and Inner Sanctums* (New York: John Wiley & Sons, 1976); Brenda Maddox, *Beyond Babel: New Directions in Communications* (New York: Simon & Schuster, 1972); and William Hickman, *Talking Moons: The Story of Communications Satellites* (New York: World Publishing Company, 1970). For a less critical view of the beginnings of Comsat as an institution and the development of international communications space technology, see Anthony Tedeschi, *Live Via Satellite: The Story of Comsat and the Technology that Changed World Communication* (Washington, DC; Acropolis Books, 1989). This work also gives a good overview of the development of Marisat and Inmarsat.

For a more detailed examination of the entire array of mobile satellite communications, see Brendan Gallagher, ed., *Never Beyond Reach: The World of Mobile Satellite Communications* (London: International Maritime Satellite Organization, 1989). The history of Comsat and the development of international space communications has also been well documented in a series of 55 oral history interviews conducted by Thomas Maxwell Safley and Nina Gilden Seavey. The interview subjects include former senior Comsat employees, members of Comsat's original board of incorporators, members of the boards of directors, former Kennedy and Johnson Administration officials, former officials from NASA, the State Department, the FCC, and congressional staffs. These holdings are housed both at the Comsat Corporation archive and at the National Air & Space Museum Department of Space History.

A comprehensive history of Intelsat has never been written, but there have been numerous studies examining a variety of aspects of the institution's development. Among these are: Marcellus Snow, *The International Telecommunication Satellite Organization: Economic and Institutional Challenges Facing an International Organization* (Baden-Baden: Nomos Verlagsgesellschaft, 1987), and Joel Alper and Joseph Pelton, *The Intelsat Global Satellite System* (New York: The American Institute of Aeronautics & Astronautics, 1984). Moreover, for the organization's 20th and 25th anniversaries, the public relations department produced short overviews of Intelsat's development. Both Comsat and Intelsat have infrequently published pocket guidebooks and charts describing the satellites as the technology has progressed. In 1989 Comsat published an extensive Satellite Reference Book available through its Corporate Affairs Office. Finally, the annual reports of Comsat, Comsat Laboratories, and Intelsat are replete with information concerning technological innovations and development as well as institutional structure and growth.

In the area of regional systems, there is not a large literature on the subject of Intersputnik because of a dearth of substantial evidence. Some basic articles worth reviewing, however, are: Simon Baker, "Soviet Communications Satellites: A Long Way to Go," *Via Satellite,* June 1990, pp. 36–38; Theo Pirard, "Intersputnik: The Eastern Brother of Intelsat," *Satellite Communications,* August 1982, pp. 38–44; "Russian Satellites and the Intersputnik System," *Space Communication and Broadcasting,* June 1988, pp. 31–35; "Closing the Credibility Gap: As CNN Signs for a Statsionar and Maxwell Gambles on a Gorizont Soviet Satellite Start to Come in From the Cold," *Cable and Satellite Europe,* September 1989, pp. 22–24. The literature on Arabsat is somewhat fuller. Some basic articles on the system include: Simon Baker, "The Tale of Arabsat," *Cable and Satellite Europe,* January 1989, pp. 59–63; and "Arabsat," *Space Communication and Broadcasting,* June 1988, pp. 65–69. Finally, for recent background on Eutelsat, articles of interest include: Simon Baker, "Bouncing Back: The Competition Grows, but So Does Eutelsat," *Cable and Satellite Europe,* May 1989, pp. 32–35; and Theo Pirard, "Eutelsat Weighs Its Options: As the 1990s Approach Eutelsat Prepares for Its Second Generation and Ponders New Projects," *Satellite Communications,* August 1989, pp. 16–17.

3. Domestic Communications Satellites

In "Space Applications—Growing Worldwide Systems," *Astronautics & Aeronautics,* June 1966, pp. 50–59, NASA's Leonard Jaffe outlined the full array of communications satellite and applications programs, including plans for the Applications Technology Satellite (ATS) series. Articles like Jaffe's signaled the push here and abroad for special domestic services such as the ATS experiment in educational programs for India's villages. The opening chapter of Wilbur L. Pritchard and Joseph A. Sculli's *Satellite Communications System Engineering* (Englewood Cliffs, NJ: Prentice Hall, 1986) describes the rise of domestic systems and presents tables on characteristics and performance of Intelsat 1–6, Comstar, Galaxy, Satcom, SBS, Telstar, Westar, ASC, Fordsat, GSTAR, Hughes Ku, RCA Ku, Spacenet, Palapa A and B, Arabsat, Aussat, Brasilsat, Telcom, direct broadcast satellites here (CBS, DBSC, RCA, STC, and USSB) and abroad (BS-2, L-Sat, TDF/TV-Sat, Tele-X, and Unisat), and several military and mobile satellites—DSCS-2, DSCS-3, Fleetsatcom, Marecs, and Marisat.

　　M. E. Sanchez Ruiz and R. D. Briskman, "Morelos—The Mexican National Satellite System," IAF paper no. 83–79 (New York: American Institute of Aeronautics & Astronautics, 1983), describe both the space and ground segments of this national system. M. G. K. Menon, "Insat in Perspective," AIAA paper no. 72–583 (New York: American Institute of Aeronautics & Astronautics, 1972), interprets the social and economic implications of India's national satellite system, an outgrowth of the ATS-series experiment to reach that country's hundreds of millions in the rural population. P. I. Klein and J. A. King, AIAA paper no. 72–521, describe the Amsat-Oscar B series of amateur radio satellites and the various intended regional uses. Giovanni Caprara, *The Complete Encyclopedia of Space Satellites* (New York: Portland House, 1986), includes descriptions of most domestic systems in a summary way. *Space Commercialization: Satellite Technology,* ed. F. Shahroki, J. S. Greenberg, and T. Al-Saud, Progress Series vol. 128 (Washington, DC: AIAA, 1990), covers various remote-sensing systems, the United States' Advanced Communications Technology Satellite (ACTS) program, Pakistan's proposed domsat (Paksat), the VSAT system for ministries of the PRC, recent developments in DBS in Japan, and new mobile links through Inmarsat. Needless to say, the literature in these areas is voluminous and varied in technical policy and coverage.

4. Spacecraft Launches

This topic permeates the professional literature of the 1950s and 1960s, as exemplified in the American Institute of Aeronautics & Astronautics' 1966 Annual Meeting forum on recoverable launch vehicles (see "The Next Generation of Launch Vehicles—Evolution or Bold Steps," *Astronautics & Aeronautics,* March 1967, pp. 57–65). The January 1973 AIAA assessment, "New Space Transportation Systems," ed. J. Preston Layton and Jerry Grey, presents the then-prevailing optimistic view of the rewards of space shuttle development and contains an extensive bibliography. For recent technical and policy developments see the AIAA professional workshop report of June 25-27, 1986, "Requirements for a U.S. Expendable Launch Vehicle Capability"; the March 1977 AIAA assessment, "The U.S. Civil Space Program"; the May 1988 U.S. Commerce Dept. report, "Space Commerce: An Industry Assessment"; the May 1989 AIAA policy paper, "Issues in Strategic Planning for Commercial Space Growth"; and the August 1989 Congressional Office of Technology Assessment (OTA) special report, *Round Trip to Orbit: Human Spaceflight Alternatives.* Joel S. Greenberg's paper "Key Issues Relating to the Commercial Development of Space" (available from Princeton Synergetics, 900 State Rd., Princeton, NJ, May 7, 1990) succinctly reviews the economic and insurance implications of, as the author says, "government agencies [lacking] a strategic plan for the commercial development of space."

5. Remote Sensing

Morton E. Davies and William R. Harris, *RAND Corporation's Role in the Evolution of Balloon and Satellite Observation Systems and Related U.S. Space Technology* (Santa Monica: RAND, 1988), authoritatively and concisely portrays the V-2 rocket's sudden illumination of space remote sensing for the conventional scientific world. Before the end of World War II, Project RAND's consultant Louis N. Ridenour, of the University of Pennsylvania's Nuclear Physics and Electronics Department, could imagine most of the main lines of "observation satellite" application. The progenitive Project RAND studies led directly to President Dwight D. Eisenhower's "Open Skies" statement on disarmament at the July 21, 1955, Geneva Conference (*Public Papers of the Presidents*). Publicly published photographs of Earth taken from a modified V-2 revealed the potential of satellite imaging; for example, see Milton Rosen, "25 years of Progress Toward Space Flight," *Jet Propulsion,* November 1955 (25th Anniversary edition). And RAND Corporation's Amron Katz disclosed the significance of a full decade of space-reconnaissance development in the American Rocket Society's *Astronautics* magazine between April and October 1960. The excellent bibliography presented by Davies and Harris captures the essentials of the early and fast-developing satellite reconnaissance world.

Articles by Amron Katz in *Astronautics & Aeronautics* later in the 1960s explore ramifications of satellite remote sensing outside the military sphere (see June, August, and October 1969 issues of *Astronautics & Aeronautics*).

By the early 1970s, NASA had embarked on the Earth Resources Technology Satellite (ERTS) Program ("The Earth Resources Program Jells," *Astronautics & Aeronautics*, April 1971), as authoritatively described in the September 1973 issue of *Astronautics & Aeronautics*. This became Landsat (see Pamela E. Mack, "The Politics of Technological Change: A History of Landsat," Ph.D. diss., Univ. of Pennsylvania, 1983). *Monitoring Earth's Ocean, Land, and Atmosphere from Space—Sensors, Systems and Applications*, ed. Abraham Schapf (New York: AIAA, 1985), ranges over the historical developments in weather and Earth-resources satellites, describes major systems, presents expert analysis of technical issues in further satellite developments, and is replete with references. *Commercial Opportunities in Space* and *Space Commercialization: Roles of Developing Countries—Vol. II Satellite Technologies and Earth-Oriented Applications* (Washington, D.C.: AIAA, 1988 and 1990), ed. F. Shahrokhi et al., show the diffusion of remote-sensing technology worldwide. D. James Baker, *Planet Earth: The View from Space* (Cambridge, MA: Harvard University Press, 1990), contains a detailed analysis of remote sensing of the Earth and an extensive bibliography. See also Earth Systems Sciences Committee of the NASA Advisory Council, *Earth System Science: A Closer View* (Washington, D.C.: National Aeronautics and Space Administration).

6. Navigation

Monte D. Wright, *History of Aerial Navigation to 1941* (Lawrence: University of Kansas Press, 1972), records the early development of instrument flying (including James H. Dolittle's pioneering experiments). Richard B. Kershner, "Transit Program Results" ("Data from Space" special issue of *Astronautics*, May 1961), describes the first major departure (aside from inertial systems) from Earth-bound navigation. In "Navigation Satellite for Worldwide Traffic Control," *Astronautics & Aeronautics*, December 1965, NASA's Eugene Ehrlich reviews the navigation satellite's potential for giving ships and aircraft rapid access to weather, rescue, and other vital information besides position fixes. And in "Satellite Aids for Aviation," *Astronautics & Aeronautics*, April 1968, pp. 64–70, Ehrlich summarizes NASA experiments in navigation and traffic control. *Avionics Navigation Systems* (New York: Wiley, 1969) reviews all systems in use at the end of the 1960s; the chapter "Satellite Navigation Aids" by Robert C. Duncan classifies and explains the then principal approach to satellite navigation. In "1980–2000: Raising Our Sights for Advanced Space Systems," *Astronautics & Aeronautics*, July/August 1976, Ivan Bekey and Harris Mayer give a lively and well-illustrated picture of the many practical uses of navigation satellites, from the personal navigation wrist-set and all-aircraft traffic control to locating nuclear material in transit and car-speed-limit control. In the same issue, Maj. Douglas Smith and Capt. William Criss of the U.S. Air Force describe the programs getting under way on the Navstar Global Positioning System (GPS), its fundamental development plan, and operational capabilities. In the March 1981 *Astronautics & Aeronautics,* Gerald K. O'Neill describes the concept for satellite air-traffic control that led to his Geostar system and the corporation that bears that name (see also his "Satellites Instead," *AOPA Pilot Magazine,* July 1982). The recent plans and operations of Geostar have been discussed in Lane F. Cooper, "Marriage of Data Communications to Satellite Technology," (*Via Satellite,* February 1990). The outcome of a decade of development in precision radio navigation can be seen in B. W. Parkinson and S. W. Gilbert, "NAVSTAR: Global Positioning System—Ten Years Later," and T. A. Stansell, Jr., "Civil GPS from a Future Perspective," both in *Proceedings of the IEEE,* vol. 2, October 1983.

7. Habitations in Space

In *The High Frontier: Human Colonies in Space* (New York: William Morrow, 1977), Gerard O'Neill recounts the evolution of space-station and space-colony concepts and supplies a rich recitation of references. The reader will do no better at first than to consult his popular, highly praised book. In "Space Colonization Now," *Astronautics & Aeronautics,* September 1975, Robert Salkeld likewise provides a panoramic bibliography. The space pioneers Tsiolkovsky, Goddard, and Oberth all considered fundamentals of living in space (see O'Neill). In his March 22, 1952, *Collier's* article "Crossing the Last Frontier," Wernher von Braun gave dramatic relief and panache to Noordung's wheel space station and cultivated popular understanding of the mainstream concepts of living and working in space. NASA's *The Apollo Spacecraft: A Chronology* (NASA SP-4009) records the debate about the advisability of a space station coincident with the Apollo lunar mission. In "An Integrated Space Program for the Next Generation," *Astronautics & Aeronautics,* January 1970, NASA associate administrator George E. Mueller describes the unfolding concepts for space-station habitats and laboratories. *Living and Working in Space: A History of Skylab* (NASA SP-4208) describes the first U.S. step toward a long-term habitat. The politics of moving from Skylab to the U.S. Space Station have been described by Hans Mark (ex-NASA Deputy Administrator, ex-Secretary of the Air Force) in *The Space Station: A Personal Journey* (Durham, NC: Duke University Press, 1987). In short, the habitat in space has over a century of thought behind it—from simple can-and-wheel space stations to Freeman J. Dyson's megalithic concept of distributing the mass of Jupiter in a satelloid field about the sun (*Science,* June 1, 1960, p. 1667).

8. *Materials Processing*

In the midst of the Apollo developments, NASA Administrator James E. Webb was preaching "commercial use of space research and technology" (*Astronautics & Aeronautics*, June 1964, pp. 74–77). In Skylab planning, this research took a strong materials bent. The May 1975 *Astronautics & Aeronautics* bannered "Materials Processing in Space—New Challenges for Industry," a lengthy article in which James H. Bredt and Brian O. Montgomery of NASA recounted results from two dozen Skylab experiments in making and modifying materials under microgravity, and presented the outlook for further experimentation, including electrophoresis to separate biologicals; they announced the availability of a bibliography at that time already containing over 500 references. Only a few years later Gerard K. O'Neill of Princeton University could describe his space colonies and mining of the moon and asteroids for large-scale space manufacturing operations (*Astronautics & Aeronautics*, March 1978, pp. 18–32). Ten years later commercial development of materials in space was "waiting for Lefty," having experienced a heavy douse of cold water from the National Research Council ("Materials Processing in Space," Report of the Space Applications Board, 1978). The NRC called the prospects for microgravity research in space "few and specific, with little if any immediate industrial potential." In the meantime, not dismayed, the Space Studies Institute had been running a biennial series of meetings on space manufacturing, the proceedings published by the American Institute of Aeronautics & Astronautics (see, for example, *Space Manufacturing 7: Space Resources to Improve Life on Earth*, Washington, DC: AIAA, 1989); NASA has sponsored a university-government-industry Center for Materials for Space Structures at Case Western Reserve (NASA PAM 525) and an industry-government Space Vacuum Epitaxy Center at the University of Houston, and has fostered a wide variety of industry space materials research projects through a volunteer industry program instigated and promoted by the AIAA. Robert J. Naumann of NASA Marshall Space Flight Center has described a novel apparatus for synthesizing new materials without contamination in space, the convex wake shield ("A More-Perfect Vacuum," *Aerospace America*, March 1987)—illustrating the point that materials research in space and space manufacturing have hardly begun as fields, and with the coming of the Space Station should see a rush of pent-up developments that the shuttle cannot support. In the meantime, reflecting the tenor of the National Research Council's Space Applications Board report "Industrial Application of the Microgravity Environment" (Washington, DC: National Academy of Sciences, 1988), basic research and the cultivation of an industrial expectancy down the road have become the recommended paths. The so-called Ride Report to the NASA Administrator of August 1987 (Sally Ride, *Leadership and America's Future in Space*, NASA Headquarters) also projects the big picture. The *Microgravity Materials Science Assessment Task Force Final Report* (NASA Headquarters) of April 1987 gives the fine grain of current planning.

9. *Financing Projects in Space*

A useful reference is Jerome Simonoff, ''Financing Space Business: What Is a Satellite in Financial Terms?'' in *The Civil Uses of Space,* ed. Costa Tsipis (New York: Pergamon Press, forthcoming). Two excellent documents from the Department of Commerce are: *Commercial Space Ventures: A Financial Perspective* (Washington, D.C.: Office of Space Commerce, April 1990), and *Space Business Indicators* (Washington, D.C.: Office of Business Analysis, January 1990). The principal feature of the former is 5 case studies of companies which found financing by one means or another; these cases demonstrate that each venture is unique. The second document includes many numbers and charts associated with space business, including a listing of all launches since the launch business was commercialized. FCC filings are a ready source of cost estimates for new space ventures; they also contain some information about how the cash requirements will be met. The trade press is also a good source of such information. See also Edward Ridley Finch, Jr., and Amanda Lee Moore, *Astrobusiness: A Guide to the Commerce and Law of Outer Space* (New York: Praeger, 1985); Joel Greenberg, Carole Gaelick, and Henry Hertzfeld, *Commercial Development of Space: Financing Mechanisms,* final report for NASA (Princeton: Princeton Synergetics, Inc., March 1989); Carole Gaelick and Brian Uzzi, *A Method for Identifying Infrastructure Commercialization Candidates,* final report for NASA (Princeton: Princeton Synergetics, Inc., July 1989). For public/private partnerships see *The Global Commons: Policy for the Planet* (Landham, MD: The Aspen Institute/University Press of America, 1990) by James Harlan Cleveland, former ambassador to NATO and longtime president of the University of Hawaii and continuing contributor at The Aspen Institute.

ABBREVIATIONS

ARPA	Advanced Research Projects Agency (later called DARPA)
ASEAN	Association of South East Asia Nations
AXAF	Advanced X-ray Astronomy Facility
C band	4–6 GHz band of microwave frequencies
CCDS	Center for the Commercial Development of Space
CDSF	Commercially Developed Space Facility
CELV	Complementary expendable launch vehicle
CEPT	Conference of European Posts and Telecommunications
CNES	Centre National d'Etudes Spatiales (French National Center for Space Studies)
Comsat	Communications Satellite Corporation
CPB	Corporation for Public Broadcasting
CRT	Cathode ray tube
DARPA	Defense Advanced Research Projects Agency
DCME	Digital Circuit Multiplication Equipment
DOC	Department of Commerce
DOD	Department of Defense
DOS	Department of State
DOT	Department of Transportation
DSCS	Defense Satellite Communications System
EBU	European Broadcast Union
EHF	Extremely high frequency (30–60 GHz)
ELDO	European Launch Development Organization
ELV	Expendable launch vehicle
Envirosat	Environmental Resources Satellite
ERTS	Earth Resources Technology Satellite
ESA	European Space Agency
ESSA	Environmental Science Service Administration (NOAA's predecessor)

Eutelsat	European Telecommunications Satellite Organization
FCC	Federal Communications Commission
FDA	Food and Drug Administration
GEO	Geostationary earth orbit
GHz	Gigahertz, a unit of frequency of microwave energy equal to one billion cycles per second
GOES	Geostationary Operational Environmental Satellite
GRO	Gamma Ray Observatory
HST	Hubble Space Telescope
Inmarsat	International Maritime Satellite Organization
Insat	Indian Telecommunications Satellite
ISY	International Space Year
ITU	International Telecommunication Union (a UN agency)
JEA	Joint Endeavor Agreement, a NASA arrangement
JPL	Jet Propulsion Laboratory
Ka band	20–30 GHz band of frequencies
Ku band	11–14 GHz band of frequencies
L band	1–2 GHz band of frequencies
LEO	Low earth orbit
LES	Lincoln Experimental Satellite
Marecs	Maritime European Communications Satellite
MHz	Megahertz, a unit of frequency of microwave energy equal to one million cycles per second
NASA	National Aeronautics and Space Administration
NESDIS	NOAA Environmental Satellite Data and Information System
NOAA	National Oceanic and Atmospheric Administration
NTIA	National Telecommunications and Information Agency
OTA	Office of Technology Assessment
OTP	Office of Technology Policy, a White House office later transferred to DOC and now called NTIA
PBS	Public Broadcasting System
PTT	Ministry of Posts, Telegraph, and Telecommunications
R&D	Research and development
S band	2–3 GHz band of frequencies
SBS	Satellite Business Systems
SICORP	Spot Image Corporation (a US corporation)
SIRTF	Space Infrared Telescope Facility
SISA	SPOT Image, Societe Anonyme (SICORP's French parent)
SPOT	Système Pour l'Observation de la Terre, ou Satellite Probatoire pour l'Observation de la Terre
SSDA	Space Systems Development Agreement (NASA)
STS	Space Transportation System (Space Shuttle)
TIROS	Television and Infrared Observation Satellite
UHF	Ultra high frequency, the 300–900 MHz band

USG   United States Government
USIA   United States Information Agency
US-ISY   US Association for the ISY
VHF   Very high frequency, the 100–300 MHz band
VSAT   Very small aperture terminal
WMO   World Meteorological Organization
X band   7–8 GHz band of frequencies

LIST OF ILLUSTRATIONS

ACKNOWLEDGMENTS

I am deeply indebted to many people who have read my manuscript and made helpful suggestions and corrections. Many colleagues at the National Academy of Engineering and at NASA—especially members of the NASA Advisory Council—have made an invaluable contribution to this work. Friends at Comsat and Hughes have been generous with their time, both in providing information and reviewing material for me.

My special thanks to Linda Brobst, Jack Hannon, Maury Mechanick, Randy Nichols, Bill Schnicke, and Elizabeth Young at Cosmat; Jim Harford and John Newbauer at AIAA; Simon Ramo at TRW; Martin Collins at NASM; Rick Anthes and Bob Serafin at UCAR/NCAR; John McElroy at University of Texas, Arlington; Jim Beggs and Chet Lee at Spacehab; Rob Briskman and Ken Manning at Geostar; Leonard Jaffe at Computer Sciences Corporation; fellow board member Randolph Ware at External Tanks Corporation; Andrea Caruso at Eutelsat; Susan Reeder at Questech, Inc.; Steve Dorfman and Harold Rosen at Hughes; Courtney Stadd at the National Space Council; John Pierce at Stanford University; Burt Edelson at Johns Hopkins University; Bob Kinzie at Intelsat; Joe Pelton at the University of Colorado; Jasper Welch at SAIC; Jim Bain, Mary Anne Haren, Ken Pedersen, Jim Rose, and Barbara Stone at NASA; George Tellmann at American Mobile Satellite Corporation; David Thompson at Orbital Sciences Corporation (on whose board I serve); Linda Billings at BDM; my fellow board member Harvey Meyerson at US-ISY; Mike Collins; Leonard David at National Space Society; Scott Chase at Phillips Publishing; Jeff Benedict, Dixie Berg, Dan Crampton, Phil

Culbertson, Dan Fink, Ed Martin, Sid Metzger, Bill Pritchard, Nina Seavy, and Albert Wheelon, consultants; Arthur Clarke in Sri Lanka; and many others who have been extremely helpful.

Special thanks also to Angela von der Lippe and Susan Wallace, my editors at Harvard University Press, who in their patient way produced a book in spite of the best efforts of the author. No acknowledgment would be complete without a word of tribute to the members of my family, who put up with much for the sake of literature.

INDEX